Higher
CHEMISTRY

Practice Workbook

© 2021 Leckie

001/18022021

10 9 8 7 6 5 4

All rights reserved. No part of this publication may be reproduced, stored in a retrieval system, or transmitted in any form or by any means, electronic, mechanical, photocopying, recording or otherwise, without the prior written permission of the Publisher or a licence permitting restricted copying in the United Kingdom issued by The Copyright Licensing Agency Ltd, 5th Floor, Shackleton House, 4 Battle Bridge Lane, London, SE1 2HX.

ISBN 9780008446741

Published by
Leckie
An imprint of HarperCollins*Publishers*
Westerhill Road, Bishopbriggs, Glasgow, G64 2QT
T: 0844 576 8126   F: 0844 576 8131
leckiescotland@harpercollins.co.uk   www.leckiescotland.co.uk

HarperCollins Publishers
Macken House,, 39/40 Mayor Street Upper, Dublin 1, D01 C9W8, Ireland

Some of this material has previously been published in the following title:
9780008263614 *Higher Chemistry: Practice Question Book* by Bob Wilson

Publisher: Sarah Mitchell
Project Managers: Harley Griffiths, Lauren Murray and Fiona Watson

Special thanks to
QBS (layout and illustration); Jess White (copy-edit);
Dylan Hamilton (proofread)

Printed by Ashford Colour Press Ltd.

A CIP Catalogue record for this book is available from the British Library.

Acknowledgements

Whilst every effort has been made to trace the copyright holders, in cases where this has been unsuccessful, or if any have inadvertently been overlooked, the Publishers would gladly receive any information enabling them to rectify any error or omission at the first opportunity.

Leckie would like to thank the following copyright holders for permission to reproduce their material:

p.25, Q2: CHARLES D. WINTERS/SCIENCE PHOTO LIBRARY; p.40, Q12: MARTYN F. CHILLMAID/SCIENCE PHOTO LIBRARY; p.46, Q3: Craig Balfour; p.55, Q3a: ANDREW LAMBERT PHOTOGRAPHY/SCIENCE PHOTO LIBRARY; p.56, Q3b: ANDREW LAMBERT PHOTOGRAPHY/SCIENCE PHOTO LIBRARY; p.56, Q3c: ANDREW LAMBERT PHOTOGRAPHY/SCIENCE PHOTO LIBRARY

This book contains FSC™ certified paper and other controlled sources to ensure responsible forest management.

For more information visit: www.harpercollins.co.uk/green

To access the ebook version of this Practice Workbook visit
www.collins.co.uk/ebooks
and follow the step-by-step instructions.

CONTENTS

About this book — 5

SECTION 1 TOPIC QUESTION PRACTICE

AREA 1 CHEMICAL CHANGES AND STRUCTURE
1. Periodicity — 9
2. Structure and bonding in compounds — 16
3. Oxidising and reducing agents — 25

AREA 2 NATURE'S CHEMISTRY
Questions testing systematic carbon chemistry are included in sections 4-11
4. Alcohols — 29
5. Carboxylic acids — 33
6. Esters, fats and oils — 35
7. Soaps, detergents and emulsions — 44
8. Proteins — 48
9. Oxidation of foods — 51
10. Fragrances — 59
11. Skin care — 63

AREA 3 CHEMISTRY IN SOCIETY
12. Getting the most from reactants — 65
13. Controlling the rate — 76
14. Chemical energy — 87
15. Equilibria — 92
16. Chemical analysis — 97

SECTION 2 MIXED EXAM QUESTION PRACTICE

Multiple-choice questions — 103
Extended response questions — 111

ANSWERS Check your answers online:
www.collins.co.uk/pages/Scottish-curriculum-free-resources

About this book

This Practice Workbook has been designed to help you feel confident about your knowledge, and about exams and assessments. It is presented in two parts to provide maximum support in both understanding and exam experience.

The topic practice section contains lots of graded practice in every single topic you will meet on your course. You can use it to consolidate your learning at any point, and to revise and refresh your knowledge in the run-up to exam time. The questions get gradually more challenging to support and extend your knowledge at the same time.

The mixed practice section then gives you the chance to put that knowledge to use in a format and standard that reflects your exams. If you get stuck on a question, you can review the relevant topic section and then come back to try it again.

Good luck!

Higher CHEMISTRY

Topic Question Practice
Bob Wilson

1 Periodicity

Exercise 1A The first 20 elements

  The modern periodic table has elements arranged in such a way as to allow chemists to make accurate predictions of physical properties and chemical behaviour for any element based on its position in the periodic table. There are recurring trends and patterns in the properties of the elements as you move from one period to the next.

a State how the elements are arranged in the periodic table.

b State the connection between the electron arrangements of elements in the same group and their chemical properties.

c State the term used to describe 'recurring trends and patterns in the properties of elements'.

  The first 20 elements in the periodic table can be divided into groupings of elements which have similar bonding and structure. These are shown as areas 1–5 in the diagram.

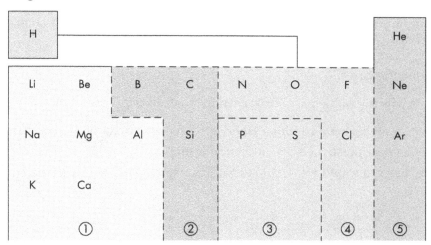

a Copy and complete the table by matching the area in the periodic table with the correct type of bonding and structure.

Area in periodic table	Bonding and structure (at normal room temperature and pressure)
	diatomic covalent gases
	solid metallic
	monatomic gases
	covalent molecular solids
	covalent network solids

b In the periodic table above, carbon is shown in area 2. It could also have been included in area 3.

Suggest what this indicates about the type of bonding and structure which can be found in carbon.

1 Periodicity   9

3 Copy and complete the table by adding the name of an element, from the first 20 in the periodic table, for each of the types of bonding and structure described.

Bonding and structure (at room temperature and pressure)	Element
metallic	
monatomic gas	
diatomic gas	
covalent molecular solid	
covalent network	

4 Copy and complete the table to show the bonding and structure of the period 3 elements in the table.

Element	Bonding	Structure
sodium		lattice
silicon		
chlorine	covalent	

5 a At one time the group 0 gases were known as the inert gases. This changed in the 1960s.

Suggest what discovery in the 1960s led to the term 'inert' no longer being used to describe the gases.

b Explain why the noble gases exist as individual atoms.

c Noble gas atoms have London dispersion forces of attraction between them.

 i Explain how these forces of attraction form.

 ii The diagram shows two noble gas atoms. Match parts A–D with the correct labels.

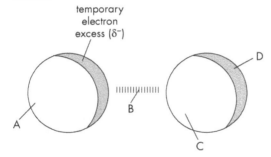

Labels: induced slight positive charge (δ^+)
temporary electron deficiency (δ^+)
induced slight negative charge (δ^-)
London dispersion force of attraction

 iii The London dispersion forces of attraction in the noble gases are only significant at low temperatures.

State what this indicates about the strength of these forces.

d Explain why the melting points and boiling points of the noble gases increase as you go down the group.

e Suggest why a traditional metal filament light bulb is filled with a mixture of argon and nitrogen instead of air.

6 a A student stated that the atoms in iodine molecules are held together by strong covalent bonds, which explains why iodine has a high melting point compared with chlorine.

Explain why this statement is not completely correct.

b Explain why fluorine has a much lower boiling point than bromine.

c i Explain why sulfur and phosphorus are solids at room temperature.

ii Explain why the melting point of sulfur is much higher than that of phosphorus.

7 Which type of structure is found in sulfur?

A Covalent network

B Covalent molecular

C Monatomic

D Metallic

8 a A form of carbon known as buckminsterfullerene is made up of C_{60} molecules. It sublimes (changes from a solid to a gas) at approximately 600°C.

Explain why buckminsterfullerene sublimes at such a high temperature for a molecular element.

b Diamond is a form of carbon which sublimes at over 3000°C.

Explain why this is so much higher than the temperature at which buckminsterfullerene sublimes.

c The diagram shows how the atoms in graphite, another form of pure carbon, are arranged.

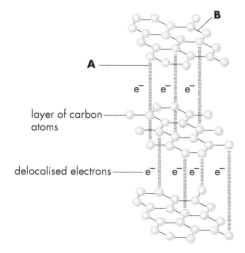

i Name the forces of attraction **A** and **B**.

ii Suggest which property graphite will have as a result of the delocalised electrons.

iii Suggest why graphite is so soft it can be used as a pencil but diamond is so hard it is used to cut glass.

1 Periodicity 11

9 The table shows the melting and boiling points of boron and silicon.

Element	Melting point (°C)	Boiling point (°C)
boron	2075	1414
silicon	4000	3265

State what the melting points and boiling points of boron and silicon indicate about the bonding and structure of the two elements.

Exercise 1B Trends in the periodic table

1 The size of an atom can be measured as the covalent radius.

   a State what is meant by the *covalent radius*.

   b The atomic number and covalent radius of some of the group 1 elements are shown in the table.

Element	Li	Na	K	Rb	Cs
Atomic number	3	11	19	37	55
Covalent radius (pm*)	134	154	196	216	235

*pm = picometre: 1 pm = 10^{-12} m

   i Draw a line graph of atomic number against covalent radius.

   ii Describe the trend in covalent radius down group 1.

   iii Explain the trend in covalent radius down a group.

2 The bar chart shows how the covalent radius of the period 2 elements changes across the period.

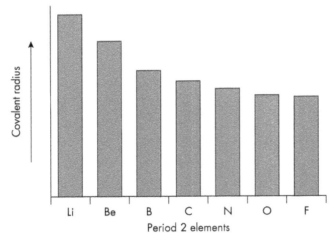

   a Describe the trend in covalent radius across period 2.

   b Explain the trend in covalent radius across a period.

   c Suggest why the covalent radius of the noble gases can't be measured.

3 a The first ionisation energy is the energy needed to remove one electron from every atom in a mole of atoms in the gaseous state.

Which of the following equations represents the first ionisation energy of chlorine?

A $Cl(g) + e^- \rightarrow Cl^-(g)$

B $Cl_2(g) + 2e^- \rightarrow 2Cl^-(g)$

C $Cl(g) \rightarrow Cl^+(g) + e^-$

D $Cl_2(g) \rightarrow 2Cl^+(g) + 2e^-$

b The table shows the first ionisation energies (in kJ mol^{-1}) of the elements in periods 2 to 5.

Li	Be	B	C	N	O	F	Ne
526	905	807	1090	1410	1320	1690	2090
Na	Mg	Al	Si	P	S	Cl	Ar
502	744	584	792	1020	1010	1260	1530
K	Ca	Ga	Ge	As	Se	Br	Kr
425	596	577	762	953	941	1150	1350
Rb	Sr	In	Sn	Sh	Te	I	Xe
409	556	556	715	816	870	1020	1170

i Ionisation energies have a positive value.

State what this indicates.

ii Describe the general trend in ionisation energies across the periods.

iii Explain the general trend in ionisation energies across the periods.

c i Describe the general trend in ionisation energies down the groups.

ii Explain the general trend in ionisation energies down the groups.

4 The table shows some ionisation energy values (in kJ mol^{-1}) for calcium forming ions.

First	Second	Third
590	1145	4912

a i The second ionisation energy of calcium is the amount of energy needed to remove a mole of electrons from a mole of gaseous calcium ions with a charge of 1+.

Write an equation to represent this process.

ii Explain why the second ionisation energy of calcium is much higher than the first.

b Suggest why it is not possible for calcium to form a Ca^{3+} ion when it reacts.

c Calculate the total amount of energy required to remove 2 moles of electrons from 1 mole of calcium atoms.

5 The bar chart shows eight successive ionisation energies for an element.

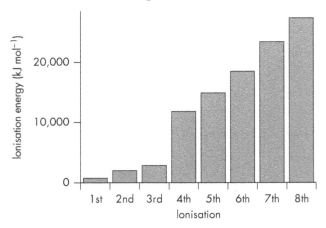

a Suggest in which group of the periodic table the element would be found.
b Explain your answer to part **a**.

6 Atoms of different elements have different electronegativities. Some of these are shown in the table.

		Group						
		1	2	3	4	5	6	7
Period	2	Li	Be	B	C	N	O	F
		1.0	1.5	2.0	2.5	3.0	3.5	4.0
	3	Na	Mg	Al	Si	P	S	Cl
		0.9	1.2	1.5	1.8	2.1	2.5	3.0
	4	K	Ca	Ga	Ge	As	Se	Br
		0.8	1.0	1.6	1.8	2.0	2.4	2.8
	5	Rb	Sr	In	Sn	Sb	Te	I
		0.8	1.0	1.7	1.8	1.9	2.1	2.5
	6	Cs	Ba	Tl	Pb	Bi	Po	At
		0.7	0.9	1.8	1.8	1.9	2.0	2.2

a Explain what is meant by *electronegativity*.
b i Describe the general trend in electronegativity across period 3 of the periodic table.
  ii Explain the general trend in electronegativity across the periodic table.
c i Describe the general trend in electronegativity down group 6 of the periodic table.
  ii Explain the general trend in electronegativity down the periodic table.
d i State which element has the highest electronegativity value.
  ii State what having a high electronegativity value indicates.
e i State which element has the lowest electronegativity value.
  ii State what having a low electronegativity value indicates.
f Suggest why there are no electronegativity values assigned to the noble gases.

7 Which of the following atoms has the greatest attraction for bonding electrons?
  A Sulfur
  B Silicon
  C Nitrogen
  D Oxygen

8 Which of the following atoms has the least attraction for bonding electrons?
  A Boron
  B Carbon
  C Aluminium
  D Silicon

9 The elements lithium to neon make up the second period of the periodic table.
  Li   Be   B   C   N   O   F   Ne

  a Name the element from period 2 which can exist as both a covalent network and a covalent molecule.
  b Explain why the first ionisation energy increases across the period.
  c Write an equation for the second ionisation energy of beryllium.
  d The atoms of which element are considered to have no bonding electrons?

2 Structure and bonding in compounds

Exercise 2A Types of chemical bond

1
a State why atoms form bonds with other atoms.
b i Name the type of bond formed between two fluorine atoms.
  ii Describe how the bond is formed between two fluorine atoms. Include a 'dot and cross' diagram.
  iii State how a covalent bond holds the atoms together.
c i Name the type of bond found in sodium chloride.
  ii Describe how the bonds in sodium chloride are formed.
d Electron density maps obtained from X-ray crystallography show areas of electron density in compounds. Electron density maps for fluorine and sodium chloride are shown.

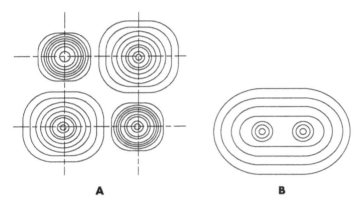

A   **B**

  i State which electron density map, **A** or **B**, is for fluorine and which is for sodium chloride.
  ii Explain your answer to part i.

2 The bonding between two hydrogen atoms is said to be pure covalent while the bonding between hydrogen and chlorine in hydrogen chloride is said to be polar covalent.

a Explain the difference between the two types of covalent bonding described.
b Draw the structural formula of hydrogen chloride and use the appropriate notation to indicate the polarity.

3 Which line in the table is correct for the polar covalent bond in hydrogen fluoride?

	Relative position of bonding electrons	Dipole notation
A	H:—F	$H^{\delta-} — F^{\delta+}$
B	H:—F	$H^{\delta+} — F^{\delta-}$
C	H—:F	$H^{\delta-} — F^{\delta+}$
D	H—:F	$H^{\delta+} — F^{\delta-}$

4 a

		Group						
		1	2	3	4	5	6	7
Period	2	Li	Be	B	C	N	O	F
		1·0	1·5	2·0	2·5	3·0	3·5	4·0
	3	Na	Mg	Al	Si	P	S	Cl
		0·9	1·2	1·5	1·8	2·1	2·5	3·0
	4	K	Ca	Ga	Ge	As	Se	Br
		0·8	1·0	1·6	1·8	2·0	2·4	2·8
	5	Rb	Sr	In	Sn	Sb	Te	I
		0·8	1·0	1·7	1·8	1·9	2·1	2·5
	6	Cs	Ba	Tl	Pb	Bi	Po	At
		0·7	0·9	1·8	1·8	1·9	2·0	2·2

Hydrogen has an electronegativity value of 2·1. Use this value and the values in the table above to predict the type of bonding found in the following substances:

   i chlorine
   ii hydrogen bromide
   iii caesium fluoride
   iv nitrogen chloride
   v sodium oxide
   vi tin(IV) chloride

> **Hint** The difference in the electronegativity values of the elements in a substance gives an indication of the bonding involved.
>
> Zero difference: pure covalent
>
> Above zero but two or less: polar covalent
>
> Above two: ionic

b i Which substance in part **a** shows the greatest ionic character?

   ii Explain your answer to part **i**.

c i Which substance in part **a** is the most polar covalent?

   ii Explain your answer to part **i**.

5 Which of the following bonds is the most polar?

A C—I

B C—F

C C—Cl

D C—Br

6 Hydrogen will form a non-polar covalent bond with an element which has an electronegativity value of:

A 0·8

B 1·6

C 2·6

D 2·2

7 The diagram represents the bonding continuum.

| 3·0 | 2·0 | 1·5 | 1·0 | 0·0 |

a Explain what is meant by the bonding continuum.
b State what the numbers represent.
c Copy the diagram and mark on it the position of substances with the following bond type:
  i pure covalent   ii ionic   iii polar covalent

Exercise 2B Intermolecular forces

1 a State the name given to intermolecular forces of attraction acting between molecules.
b i Name **three** types of intermolecular force acting between molecules.
  ii State which of the forces in part i are present in all substances.

2 Explain why all covalent molecular substances can exist as solids, liquids or gases.

3 The table below shows the boiling points of some alkanes.

Alkane	Molecular mass (g)	Boiling point (°C)
methane	16	−162
ethane	30	−89
propane	44	−42
butane	58	0
pentane	62	36

a i Alkane molecules are described as non-polar.
    Explain what is meant by the term *non-polar*.
  ii Name the force of attraction which exists between alkane molecules.
  iii Explain how the force of attraction in part **ii** is formed.
b i Describe the trend in the boiling points of the alkanes as the molecular mass increases.
  ii Explain the trend in the boiling points of the alkanes.

4 The structural formulae for some compounds which contain polar bonds are shown below.

A   H — Cl

B   O = C = O

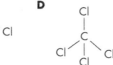

a Identify which of the molecules are polar and which are non-polar.
b Explain your answers to part **a**.
c Name the main type of attraction which exists between polar covalent molecules.
d Name the type of attraction which exists between non-polar covalent molecules.

e When a charged rod was brought close to a thin stream of liquid, the stream of liquid was attracted to the charged rod.
  i State whether the liquid was polar or non-polar.
  ii Explain your answer to part **i**.

5 The diagram below shows three trichloromethane molecules.

  a i State whether a trichloromethane molecule is polar or non-polar.
   ii Explain your answer to part **i**.
  b i Name the main force of attraction which exists between the molecules.
   ii Draw the three molecules and add broken lines to show the force of attraction named in part **i**.

6 The table gives some information about bromine (Br_2) and iodine monochloride (ICl).

Compound	Number of electrons	Melting point (°C)
bromine	70	−7
iodine monochloride	70	+27

Explain why the melting points of the two compounds are so different even though the number of electrons in each molecule is the same.

7 The diagram shows the forces of attraction, **X** and **Y**, between two molecules of 1-chlorooctane.
  a Name the forces of attraction **X** and **Y**.
  b Suggest why the forces of attraction **X** are significant here.

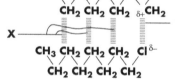

8 Which line in the table shows the correct entries for tetrachloromethane (CCl_4)?

	Polar bonds?	Polar molecules?
A	yes	yes
B	yes	no
C	no	no
D	no	yes

9 Which of the following has more than one type of van der Waals' force operating between its molecules in the liquid state?

  A   Br — Br

  B   O=C=O

  C SiH_4

  D PCl_3

10 London dispersion forces and polar covalent bonds both involve the formation of dipoles.

State what the difference is between the dipoles formed in each case.

Exercise 2C Hydrogen bonding

1 Highly polar bonds are formed when hydrogen is attached to atoms of certain elements with a high electronegativity value. Oxygen is one of these elements.

   a Name the other **two** elements.

   b Water molecules have these highly polar bonds.

   Name the type of intermolecular force which exists between water molecules.

   c The diagram shows three water molecules.

   Copy the diagram and add:

   i the symbols δ^+ and δ^- to the appropriate part of each molecule

   ii the intermolecular forces from part **b**.

2 The graph shows the boiling points of the hydrides of the elements in groups 4–7 of the periodic table.

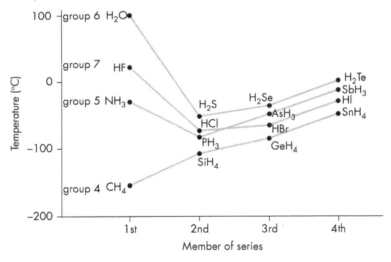

   a i Name the force of attraction between the molecules of the group 4 hydrides.

   ii Describe the trend in the boiling points of the group 4 hydrides as the molecules increase in size.

   iii Explain the trend in the boiling points of the group 4 hydrides.

   b Explain why it might be expected that water would have a boiling point well below zero but it is in fact 100°C.

   c Explain why the boiling points of ammonia (NH_3) and hydrogen fluoride (HF) are not what would have been expected if they followed the same pattern as the other hydrides in their respective groups.

3 The graph shows the melting points of the hydrides of the elements in groups 4–7 of the periodic table.

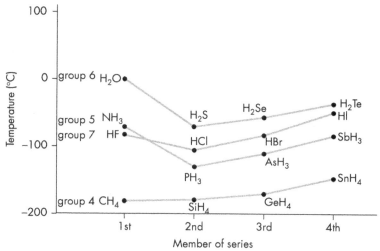

   a i Name the force of attraction between the molecules of the group 4 hydrides.
      ii Describe the trend in the melting points of the group 4 hydrides as the molecules increase in size.
      iii Explain the trend in the melting points of the group 4 hydrides.
   b Explain why the melting points of water (H_2O), ammonia (NH_3) and hydrogen fluoride (HF) are not what would have been expected if they followed the same pattern as the other hydrides in their respective groups.

4 Solid substances normally sink when put into their liquid form. For example, a solid piece of wax sinks in liquid (melted) wax. Solid water (ice), however, is unusual in that it floats on its liquid form (water).

Explain this unusual property of water.

5 Hydrogels have the ability to absorb great amounts of water. The diagram below shows part of the structure of a hydrogel.

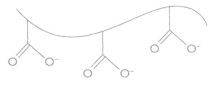

   a State the type of bonding that exists between the hydrogel and water molecules.
   b Copy the diagram and show how three molecules of water attach themselves to the surface of the hydrogel.
   c Hydrogels can be added to water to improve water's efficiency in terms of putting out fires. The water/hydrogel sticks to the surface of the material which is burning.

   This will not work if seawater is used instead of freshwater.

   Suggest a reason for this observation.

6 Which of the following is not a van der Waals' force?

   A Covalent bond
   B Hydrogen bond
   C London dispersion force
   D Permanent dipole–permanent dipole attraction

2 Structure and bonding in compounds

Exercise 2D Viscosity, solubility and miscibility

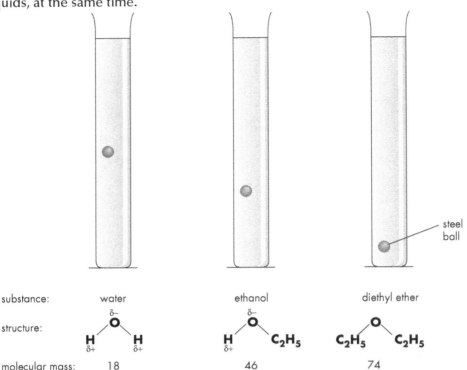

1. The diagram shows what happens to steel balls when they are dropped into different liquids, at the same time.

   a When looking at the molecular mass of the compounds, it might be expected that diethyl ether would be the most viscous.

   Explain why this is not the case.

   b Explain why water is more viscous than ethanol.

   c The experiment was repeated using two alcohols, propane-1,2,3-triol and propane-1,2-diol.

   The steel ball fell faster in propane-1,2-diol.

   i State what this result tells us about the viscosity of the two alcohols.

   ii Explain why there is a difference in the viscosity of the two alcohols.

2. Water is sometimes called the 'universal solvent'.

   a Suggest why this is the case.

   b Hydrogen chloride can be represented as $H^{\delta+}-Cl^{\delta-}$.

   Explain why hydrogen chloride is soluble in water but not in hexane.

c The simplified structure of a glucose molecule is shown.

Explain why glucose is able to dissolve in water.

d i Explain why sodium chloride is soluble in water but not in hexane.

  ii Draw a diagram to show what happens to the Na^+ and Cl^- ions when sodium chloride dissolves in water. Show four water molecules in each case.

e Suggest why the natural oils from the skin which trap dirt on clothes don't dissolve in water but will dissolve in tetrachloromethane.

3 a Water and ethanol are miscible and this forms the basis for alcoholic drinks.

The structure of an ethanol molecule is shown.

Explain, with the aid of a diagram, why ethanol and water are miscible.

b Crude oil is mainly a mixture of hydrocarbons. When crude oil is spilled into the sea the oil and water form two separate layers.

Explain this observation.

c When bromine solution is shaken with pentane, the bromine quickly moves from the water to the pentane.

Explain this observation.

d Write a solubility rule for ionic, polar covalent and non-polar substances and the type of solvent they will dissolve in.

4 The diagram shows part of a protein chain and the different forces of attraction, **A–D**, which hold it in a specific shape.

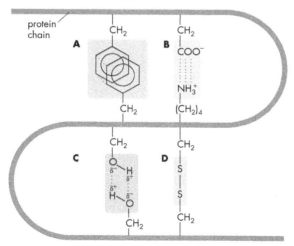

Give names for the four different forces of attraction, **A–D**.

5 Use the word bank to help you complete the summary of intermolecular forces of attraction.

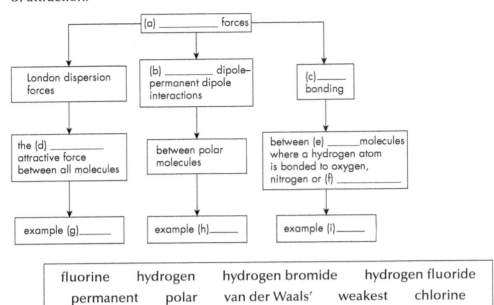

| fluorine | hydrogen | hydrogen bromide | hydrogen fluoride |
| permanent | polar | van der Waals' | weakest | chlorine |

3 Oxidising and reducing agents

> **Hint** Use the electrochemical series in the SQA data booklet when answering questions in this chapter.

Exercise 3A Elements as oxidising and reducing agents

1 In the thermite reaction, aluminium reacts with iron(III) oxide to produce molten iron and aluminium oxide. The molten iron can be run off to repair cracks in the likes of railway lines. The ion-electron equations for the reaction are:

$Al(s) \rightarrow Al^{3+}(s) + 3e^-$

$Fe^{3+}(s) + 3e^- \rightarrow Fe(\ell)$

  a **i** Identify the oxidation equation.

     ii Identify the reduction equation.

  b Combine the oxidation and reduction equations to give the redox equation.

  c **i** Identify the reducing agent.

     ii Identify the oxidising agent.

2 When zinc metal is put into a solution of copper(II) sulfate, the blue colour of the solution fades over time and becomes colourless.

Time elapsed (mins): 10   20   30

  a Suggest:

    i what is happening on the surface of the zinc

    ii why the solution changes from blue to colourless.

  b **i** Write an ion-electron equation for the reduction reaction.

     ii Write an ion-electron equation for the oxidation reaction.

  c Combine the reduction and oxidation equations to give the redox equation.

  d **i** Identify the reducing agent.

     ii Identify the oxidising agent.

3 Bromine is produced from water from inland seas like the Dead Sea, which contains a high concentration of bromide ions. When chlorine gas is bubbled through the seawater the bromide ions are displaced as bromine solution and the chlorine forms chloride ions.

   a **i** Write an ion-electron equation for the reduction reaction.

   ii Write an ion-electron equation for the oxidation reaction.

   b Combine the reduction and oxidation equations to give the redox equation.

   c **i** Identify the reducing agent.

   ii Identify the oxidising agent.

4 Copper metal reacts with silver(I) nitrate solution to form copper(II) nitrate solution and silver metal.

   a **i** Write an ion-electron equation for the reduction reaction.

   ii Write an ion-electron equation for the oxidation reaction.

   b Combine the reduction and oxidation equations to give the redox equation.

   c **i** Identify the reducing agent.

   ii Identify the oxidising agent.

5 Chromium metal reacts with nickel(II) sulfate solution to form chromium(III) sulfate solution and nickel metal.

   a **i** Write an ion-electron equation for the reduction reaction.

   ii Write an ion-electron equation for the oxidation reaction.

   b Combine the reduction and oxidation equations to give the redox equation.

   c **i** Identify the reducing agent.

   ii Identify the oxidising agent.

6 **a** **i** Write a statement linking the electronegativity value of an element and whether it is a reducing agent or an oxidising agent.

   ii Explain your answer to part **i**.

   b **i** State in which group of the periodic table the strongest reducing agents are found.

   ii State in which group of the periodic table the strongest oxidising agents are found.

Exercise 3B Compounds and group ions as oxidising and reducing agents

1 Iron can be extracted from iron(III) oxide in a blast furnace by reacting it with carbon monoxide.

$$Fe_2O_3(s) + 3CO(g) \rightarrow 2Fe(s) + 3CO_2(g)$$

a State what the carbon monoxide is acting as.
b Write the ion-electron equation for the reduction reaction.

2 Bromine solution is decolourised by sulfite ions. Bromide ions and sulfate ions are formed.
a i Write the ion-electron equation for the reduction reaction.
  ii Write the ion-electron equation for the oxidation reaction.
b Combine the reduction and oxidation equations to give the redox equation.
c i Identify the reducing agent.
  ii Identify the oxidising agent.

3 A solution of iron(II) ions reacts with acidified hydrogen peroxide (H_2O_2) solution to form iron(III) ions and water.
a i Write the ion-electron equation for the oxidation reaction.
  ii Complete the reduction ion-electron equation for hydrogen peroxide forming water:
  $$H_2O_2(aq) \rightarrow H_2O(\ell)$$
b Combine the reduction and oxidation equations to give the redox equation.
c i Identify the reducing agent.
  ii Identify the oxidising agent.

4 Sulfite ions form sulfate ions when they react with acidified hydrogen peroxide (H_2O_2) solution. The hydrogen peroxide forms water.
a i Write the ion-electron equation for the reduction reaction.
  ii Write the ion-electron equation for the oxidation reaction.
b Combine the reduction and oxidation equations to give the redox equation.
c i Identify the reducing agent.
  ii Identify the oxidising agent.

5 When acidified potassium permanganate solution ($KMnO_4(aq)$) reacts with iron(II) ions, manganese(II) ions and iron(III) ions are produced.
a i Write the ion-electron equation for the oxidation reaction.
  ii Complete the reduction ion-electron equation:
  $$MnO_4^-(aq) \rightarrow Mn^{2+}(aq)$$
b Combine the reduction and oxidation equations to give the redox equation.
c i Identify the reducing agent.
  ii Identify the oxidising agent.

6 When acidified potassium dichromate solution ($K_2Cr_2O_7$(aq)) reacts with tin(II) chloride, chromium(III) ions and tin(IV) ions are formed.

   a **i** Write the ion-electron equation for the oxidation reaction.

      ii Complete the reduction ion-electron equation:

        $Cr_2O_7^{2-}$(aq) $\rightarrow$ Cr^{3+}(aq)

   b Combine the reduction and oxidation equations to give the redox equation.

   c **i** Identify the reducing agent.

      ii Identify the oxidising agent.

7 **a** Give **one** example of how an oxidising agent can be used to treat a medical condition.

   b Give **one** example of how hydrogen peroxide is used as a bleach.

8 Some oxidising and reducing agents are listed below:

$Cr_2O_7^{2-}$(aq)   Li(s) Zn^{2+} Cr^{3+}(aq) Li^+(aq) MnO_4^-(aq)   Zn(s) Mn^{2+}(aq)

   a Select the strongest reducing agent.

   b Select the strongest oxidising agent.

9 **a** Predict whether each of the following combinations would result in a reaction.

     i Chlorine gas bubbled through a solution of iodide ions.

     ii Silver metal added to copper(II) sulfate solution.

     iii Acidified potassium permanganate solution added to tin(II) chloride solution.

     iv Chromium(III) ions mixed with bromine solution.

   b For each of the reactions which takes place in part **a**, write the ion-electron equations for the reduction and oxidation reactions.

   c For each of the reactions which takes place in part **a**, identify the oxidising and reducing agents.

4 Alcohols

Exercise 4A Naming alcohols

  The structural formulae of some straight chain alcohols are shown.

A
```
   H   H   H
   |   |   |
H—C—C—C—O—H
   |   |   |
   H   H   H
```

B
```
   H   H   H
   |   |   |
H—C—C—C—H
   |   |   |
   H   O   H
       |
       H
```

C
```
   H   H   H   H   H   H
   |   |   |   |   |   |
H—C—C—C—C—C—C—H
   |   |   |   |   |   |
   H   H   H   O   H   H
               |
               H
```

D
```
   H   H   H   H   H   H   H
   |   |   |   |   |   |   |
H—C—C—C—C—C—C—C—H
   |   |   |   |   |   |   |
   H   H   H   H   O   H   H
                   |
                   H
```

  a Name each alcohol.
  b i Write the molecular formula for each of the alcohols.
     ii Write the shortened structural formula for each of the alcohols.
  c Name the functional group in alcohols.
  d Alcohols **A** and **B** have the same molecular formula but different structures.
     i Write the molecular formula for **A** and **B** showing the functional group.
     ii State what term is used to describe molecules with the same molecular formula but different structures.
  e State whether each of the alcohols is primary, secondary or tertiary.

  a Draw structural formulae for the following alcohols.
     i   Butan-1-ol
     ii Butan-2-ol
     iii 2-methylpropan-2-ol
  b Write molecular formulae for each of the alcohols in part **a**.
  c The alcohols in part **a** are isomers.
     State what is meant by *isomers*.
  d State whether each of the alcohols is primary, secondary or tertiary.

3 The structural formulae of some branched chain alcohols are shown.

A, **B**, 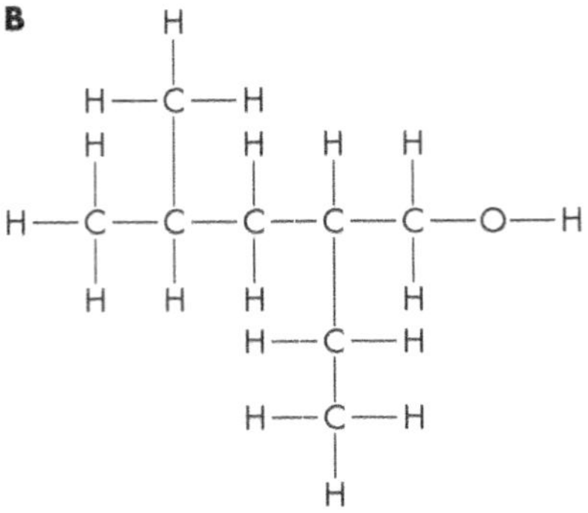 **C**, 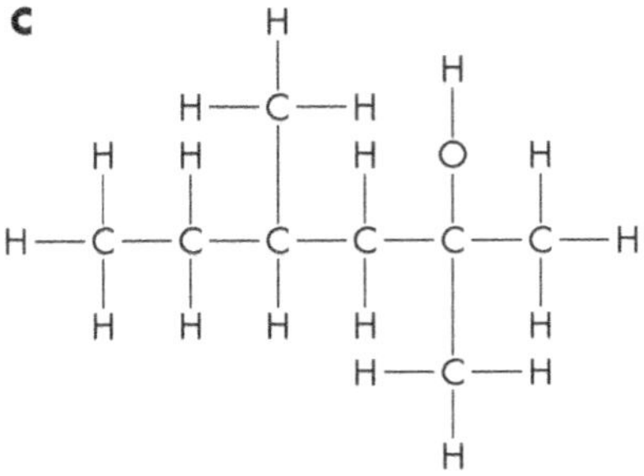 **D** (structures shown)

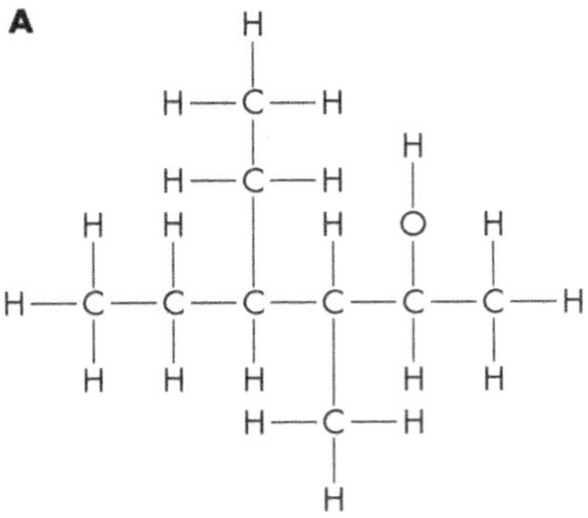

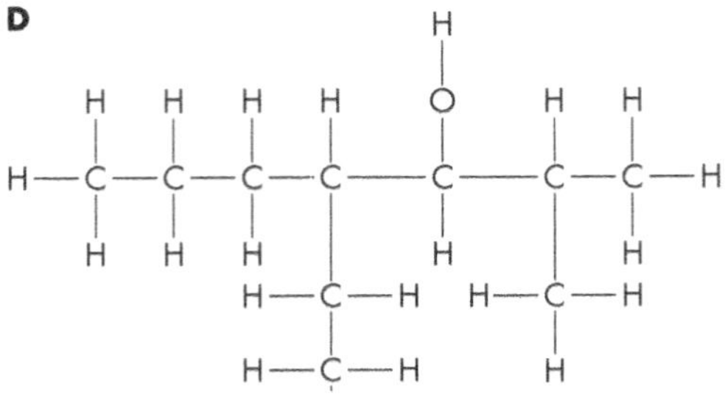

a Name each alcohol.
b i Write the molecular formula for each of the alcohols.
 ii Write the shortened structural formula for each alcohol.
c Molecules **B** and **D** have the same molecular formula but different structural formulae. State the term used to describe these types of molecules.
d State whether each of the alcohols is primary, secondary or tertiary.

4 a Draw the structural formula for each of these alcohols.
  i 2-methylpropan-1-ol
  ii 3-methylpentan-3-ol
  iii 3,5-dimethylhexan-2-ol
  iv 4-ethyl-2,6-dimethylheptan-3-ol
 b Write the molecular formula for each alcohol in part **a**.
 c Write the shortened structural formula for each alcohol in part **a**.
 d State whether each of the alcohols in part **a** is primary, secondary or tertiary.

5 a i Give the name of the type of alcohol which has two hydroxyl groups.
  ii Give the name of the type of alcohol which has three hydroxyl groups.
 b The structural formulae of some alcohols are shown.

 A, **B**, 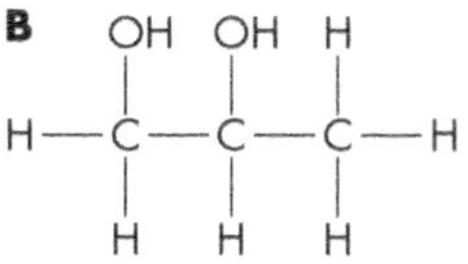 **C**, 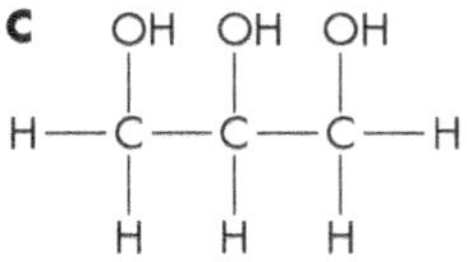 **D** (structures shown)

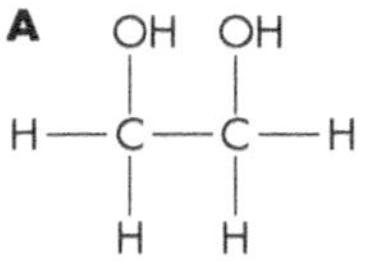

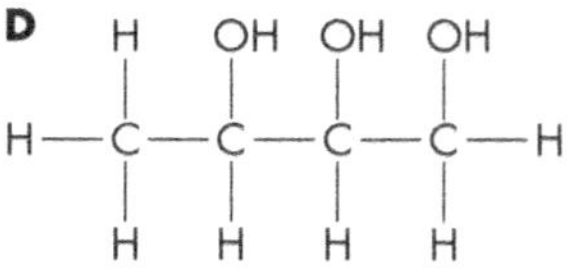

 Identify which are diols and which are triols.

Exercise 4B Properties of alcohols

1 The bar chart compares the boiling points of some primary alcohols and alkanes with the same number of carbons.

a State the trend in boiling points for the alcohols and alkanes.

b Explain why the alcohols have much higher boiling points than the corresponding alkanes.

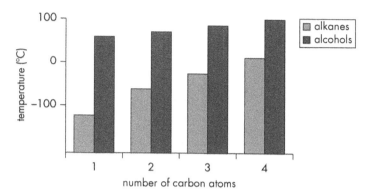

2 The table gives the boiling points of some primary and secondary alcohols.

Primary alcohol	Boiling point (°C)	Secondary alcohol	Boiling point (°C)
propan-1-ol	97	propan-2-ol	82
butan-1-ol	116	butan-2-ol	100
pentan-1-ol	137	pentan-2-ol	118

a Write a general statement about the boiling points of primary alcohols compared with secondary alcohols.

b Explain the reason for the difference in boiling points between the two types of alcohol.

3 a Explain why methanol, ethanol and propan-1-ol are totally miscible with water.

b The table shows the solubility of some primary alcohols.

Alcohol	Formula	Solubility in water (g l⁻¹)
butan-1-ol	C_4H_9OH	63.0
pentan-1-ol	$C_5H_{11}OH$	22.0
hexan-1-ol	$C_6H_{13}OH$	5.9
heptan-1-ol	$C_7H_{15}OH$	1.7
octan-1-ol	$C_8H_{17}OH$	0.5

i State the trend in solubility of the alcohols in the table.

ii Explain the trend in solubility of the alcohols in the table.

4 **a** The table compares the boiling points of some alcohols and fluoroalkanes.

Alcohols	Boiling point (°C)	Fluoroalkanes	Boiling point (°C)
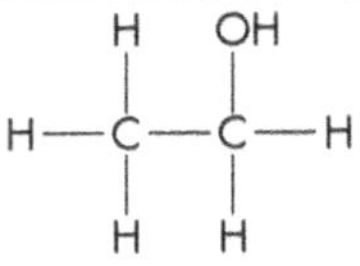 ethanol	78	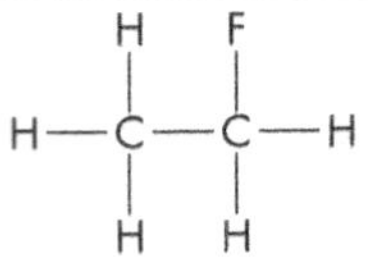 fluoroethane	−37
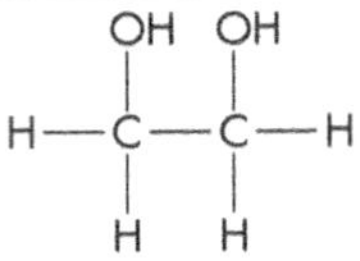 ethane-1,2-diol	196	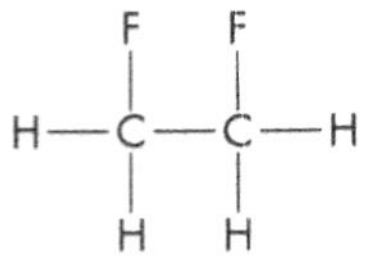 1,2-difluoroethane	31
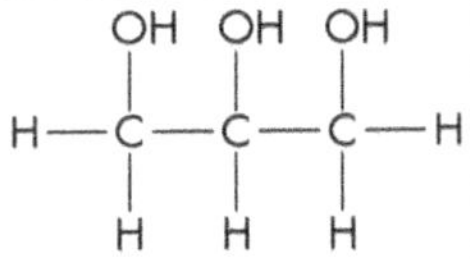 propane-1,2,3-triol	290	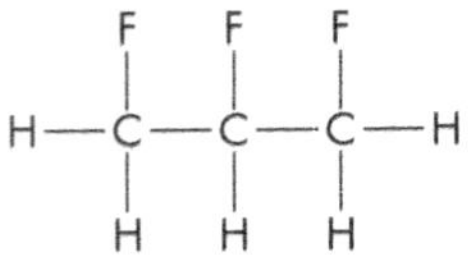 1,2,3-trifluoropropane	71

Explain why alcohols have higher boiling points than their equivalent fluoroalkane even though they have the same number of electrons.

b The structures of propan-1-ol and ethane-1,2-diol are shown.

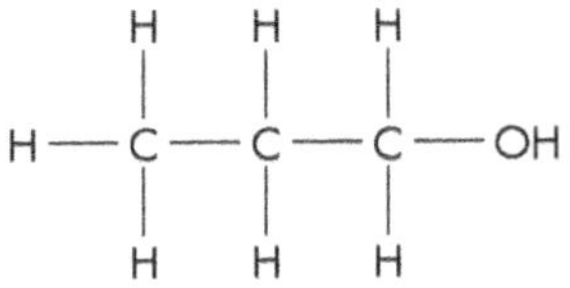

propan-1-ol b.p. = 97°C

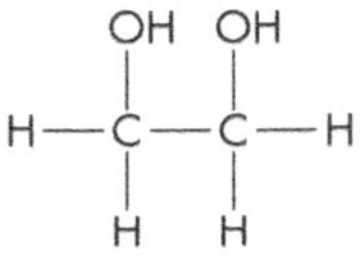

ethane-1,2-diol b.p. = 196°C

Explain why there is such a big difference in the boiling points of the two alcohols.

c Explain why diols and triols are more soluble in water than alcohols with only a single hydroxyl group.

d i Arrange the following compounds in order of how soluble they are in water, with the least soluble first.

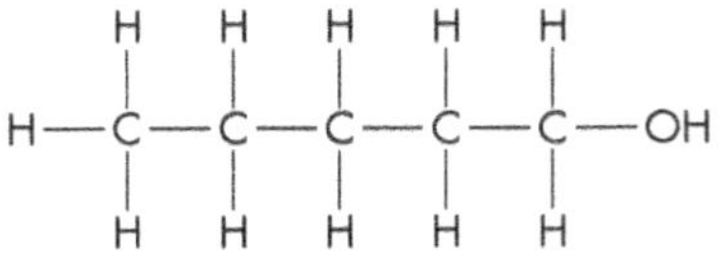

pentan-1-ol

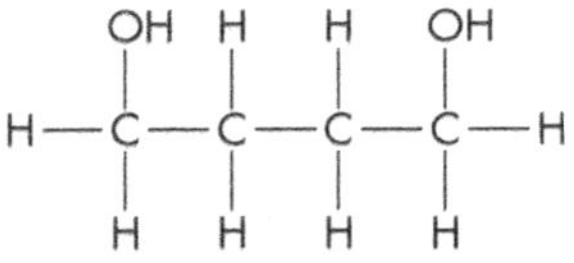

butane-1,4-diol

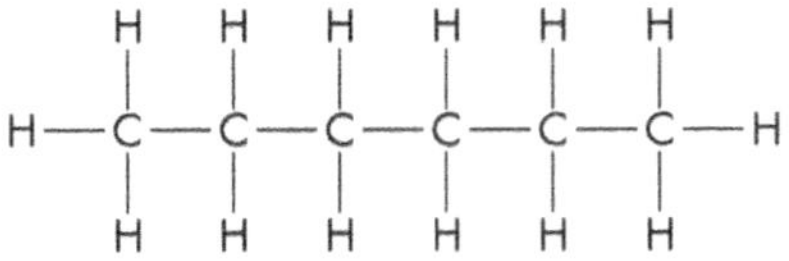

hexane

ii Explain your answer to part **i**.

e Explain why the more hydroxyl groups an alcohol molecule has, the more viscous the alcohol is.

5 Carboxylic acids

Exercise 5A Naming and reactions of carboxylic acids

1 The structural formulae of some straight chain carboxylic acids are shown.

a Name each carboxylic acid.
b Write the molecular formula for each of the acids in part **a**.
c Write the shortened structural formula for each of the acids in part **a**.
d Name the functional group in carboxylic acids.

2 The structural formulae of some branched chain carboxylic acids are shown.

a Name each carboxylic acid.
b Write the shortened structural formula for each carboxylic acid.

3 a Draw the structural formula for each of these carboxylic acids.
  i 3-methylpentanoic acid
  ii 3,4-dimethylhexanoic acid
  iii 3-ethyl-2-methylheptanoic acid
  iv 2-ethyl-2,4-dimethylpentanoic acid
 b Write the shortened structural formula for each carboxylic acid in part **a**.
 c Write the molecular formula for each carboxylic acid in part **a**.

4 a Complete the following word equations.
  i ethanoic acid + sodium hydroxide →
  ii propanoic acid + calcium hydroxide →
  iii ethanoic acid + calcium carbonate →
  iv propanoic acid + magnesium carbonate →
 b Write balanced chemical equations for each of the reactions in part **a**.
 c State what type of reaction is taking place when a carboxylic acid reacts with a base.
 d Suggest why a kitchen worktop made from marble (calcium carbonate) can be damaged if vinegar is spilled on its surface.
 e Sodium benzoate is used as a food preservative. When it dissolves, the benzoate ions combine with hydrogen ions to form the acid.
  i Name the acid formed from sodium benzoate.
  ii Suggest why this process works well in fizzy drinks and foods containing citrus fruits.

6 Esters, fats and oils

Exercise 6A Esters

1. Which of the following is the name of an ester?
   A ethanal
   B ethanol
   C methyl ethanoate
   D sodium ethanoate

2. Which of the following is the structural formula of an ester?

A
```
         H
         |
    H—C—H   H
         |   |
H—O—C———C—H
         |   |
         H   H
```

B
```
       O
      ‖
  H—C
      \
       O—H
```

C
```
    H   O       H   H   H
    |   ‖       |   |   |
H—C—C—O—C—C—C—H
    |           |   |   |
    H           H   H   H
```

D
```
    H   O   H   H
    |   ‖   |   |
H—C—C—C—C—H
    |       |   |
    H       H   H
```

3. Which of the following is an ester link?

A
```
  O
  ‖
—C—H
```

B
```
  O
  ‖
—C—O—H
```

C
```
  |
—C—O—H
  |
```

D
```
  O
  ‖
—C—O—
```

6 Esters, fats and oils

4 The table shows some alcohols and carboxylic acids.

	Alcohol	Carboxylic acid
i	methanol	ethanoic acid
ii	ethanol	propanoic acid
iii	propan-1-ol	methanoic acid
iv	methanol	propanoic acid
v	ethanol	ethanoic acid
vi	propan-1-ol	ethanoic acid

a Name the ester formed when each pair of alcohol and carboxylic acid reacts.
b Draw the structural formula of each of the esters formed in part **a**.
c i Name the type of reaction which takes place when esters are made.
  ii Name the small molecule which is also formed when esters are produced.

5 a Name each of the esters from the structural formulae shown.

A

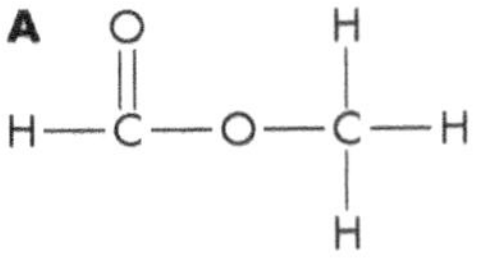

B

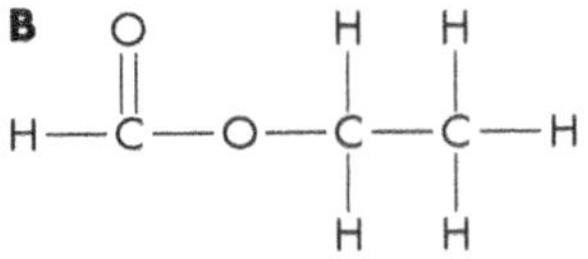

C

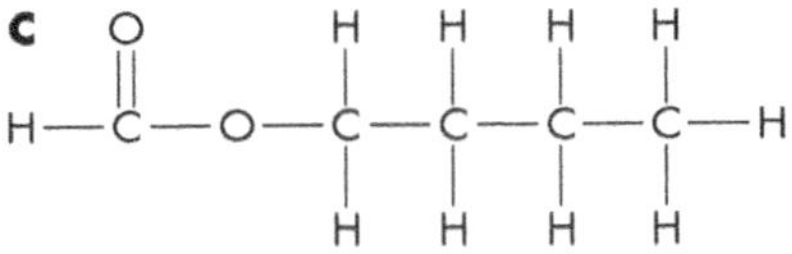

D

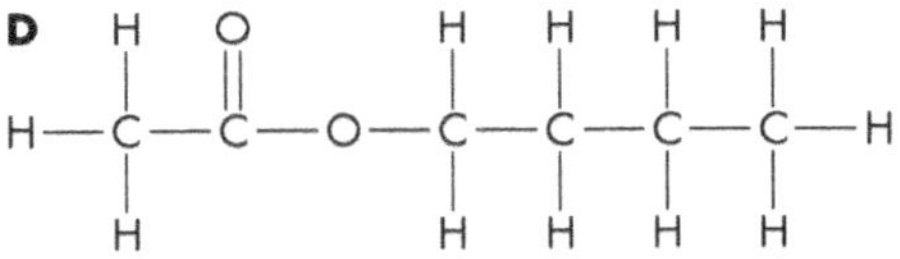

E

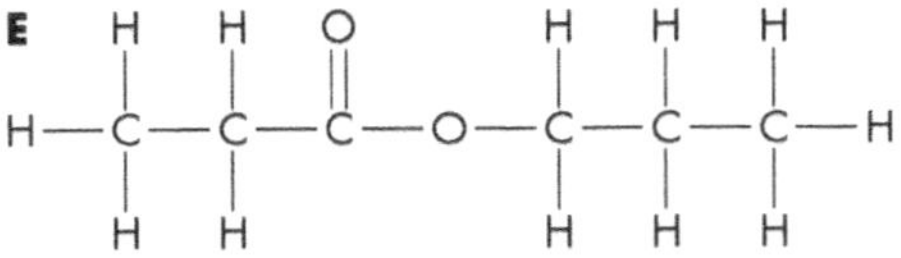

F

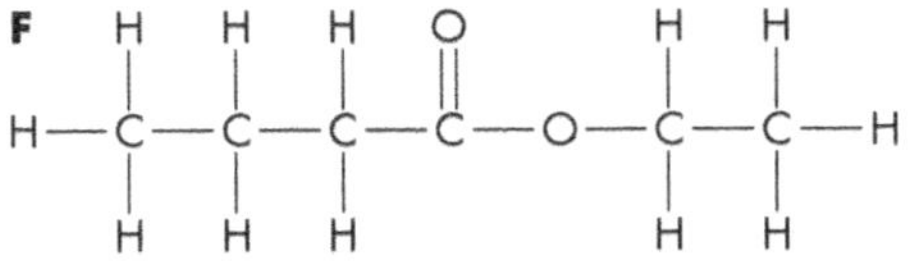

b Name the alcohol and carboxylic acid from which each of the esters in part **a** is made.

6 a Draw the structural formula of each of the following esters.
  i Methyl butanoate
  ii Butyl propanoate
  iii Butyl butanoate
  iv Propyl butanoate

b Write the molecular formula of each of the esters in part **a**.
c i State which of the two esters in part **a** are isomers.
  ii Explain your answer to part **i**.

7 Name the parent alcohol and carboxylic acid from the structural formulae of the following esters.

a

b

c

8 The structural formulae of an alcohol and a carboxylic acid are shown.

a Name the alcohol and carboxylic acid.
b Draw out the structures and show, by drawing a circle around the parts of the two functional groups which react, how the ester is formed.
c i Draw the structural formula of the ester.
  ii Write the molecular formula for the ester.
  iii Name the ester.
  iv Name the small molecule which is formed.
d i Write a word equation for the reaction between the carboxylic acid and alcohol.
  ii Name the type of reaction taking place.

9 Ethyl ethanoate can be made in the laboratory as shown.

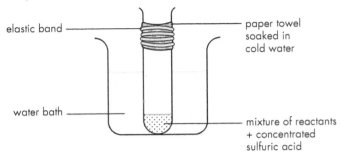

> **Area 4** – Researching Chemistry:
> General practical techniques
>
> **a i** Give a reason for using a water bath for heating the mixture.

  ii State the purpose of the paper towel soaked in cold water.
  iii State how you would know that the ester had been formed.
  iv Name the type of reaction taking place.
 b i Draw the structural formula for ethyl ethanoate.
  ii Name the parent alcohol and carboxylic acid from which ethyl ethanoate is made.
  iii Draw structural formulae for the alcohol and carboxylic acid from which ethyl ethanoate is made.

10 a Esters can be used as flavourings.

Give **two** other uses for esters.

 b i The table shows some esters which are used as flavourings. Copy and complete the table.

Flavouring	Ester	Parent alcohol	Parent carboxylic acid
pineapple		propan-1-ol	pentanoic acid
apple	ethyl butanoate		
orange		octan-1-ol	ethanoic acid
banana	pentyl ethanoate		

  ii Draw the structural formulae for each of the esters in the table.
 c Ethyl ethanoate is found in fruits and other foods and is also made industrially on a large scale. One of its uses is to extract the caffeine from coffee and tea.
  i Suggest what ethyl ethanoate is acting as when it extracts caffeine.
  ii Suggest why companies who use this technique can claim that their products are naturally decaffeinated.

iii Vitamin C is also found in fruits. The structure of vitamin C is shown below.

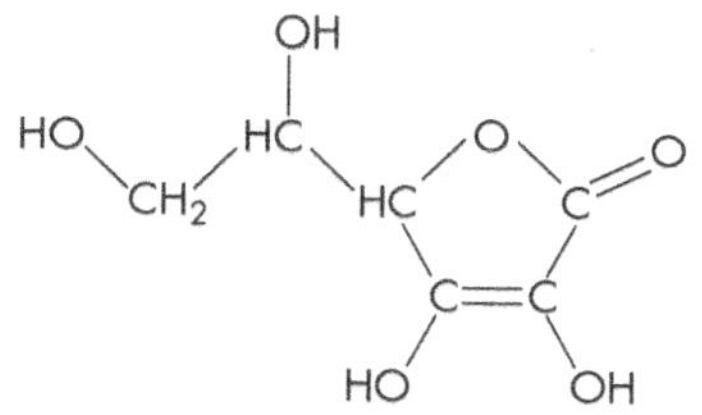

Draw the structure and circle the ester group.

iv Write the molecular formula for vitamin C.

11 Esters are common solvents. They are used in products such as nail varnish and spray paints and have a distinctive smell. They are classed as volatile organic compounds (VOCs).

a i What does the information above suggest about the boiling points of esters.

ii Explain your answer to part **i**.

b Suggest why their volatility is a useful property when esters are used in this way.

c Suggest why it is recommended that the user wears a mask and works in a well-ventilated area when using spray paint.

d Esters contribute to the flavour of drinks such as wine.

Suggest why a wine maker swirls wine in a glass and smells it when testing its flavour.

e Suggest why a perfume loses its smell after a few hours.

12 Esters can be broken down into their parent alcohol and carboxylic acid.

a Give the name of this process.

b i Name the alcohol and carboxylic acid produced when methyl propanoate is broken down by warming it with dilute hydrochloric acid.

ii Draw structural formulae for methyl propanoate and the alcohol and carboxylic acid formed.

c The structural formula of an ester is shown.

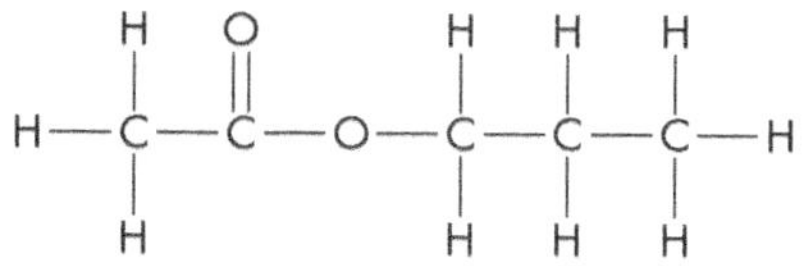

i Name the ester.

ii Draw the structural formulae of the products formed when the ester is broken down using dilute hydrochloric acid and name the products.

d Esters can be broken down using dilute alkali such as sodium hydroxide.

ethyl ethanoate + sodium hydroxide → sodium ethanoate + ethanol

i What type of compound is sodium ethanoate?

ii Explain why sodium ethanoate is formed and not ethanoic acid.

6 Esters, fats and oils

e Esters can be broken down using the apparatus shown.

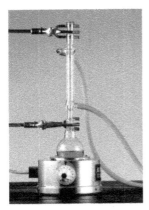

> **Area 4 –** Researching Chemistry:
> Common chemical apparatus
> General practical techniques

   i Suggest why a condenser has to be used.
   ii State why an electric heater is used instead of a Bunsen burner.

13 Explain the following observations.

a An ester is added to distilled water. After an hour the mixture has a pH of less than 7.

b The fruity fragrances in an old bottle of perfume are replaced by an unpleasant smell.

Exercise 6B Fats and oils

1 The structures of two molecules found in fats and oils are shown.

A H_2C-OH
     $HC-OH$
     H_2C-OH

B $H-O$, CH_2 CH_2 CH_2 CH_2 CH_2 CH_2 CH_2 CH_2 CH_3
       C CH_2 CH_2 CH_2 CH_2 CH_2 CH_2 CH_2 CH_2
       O

a State the family of compound that fats and oils belong to.
b Name **A**.
c Name the type of compound **B** is.
d i Name the type of reaction which takes place when molecules **A** and **B** react.
  ii State how many molecules of **B** would be needed to react completely with one molecule of **A**.
  iii Explain your answer to part **ii**.

2 Esters found in fats and oils can be represented as shown below.

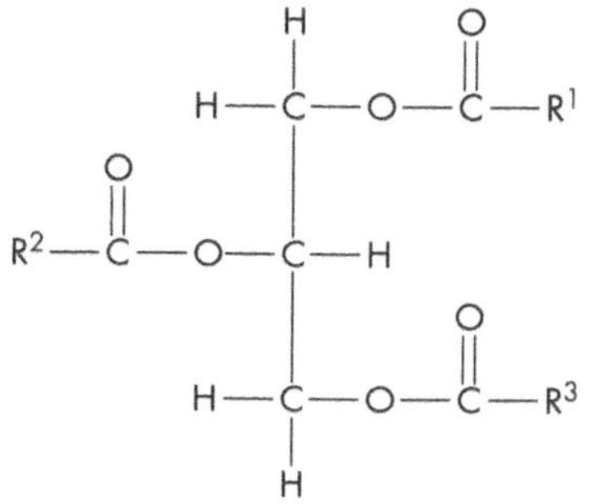

$$
\begin{array}{c}
\overset{H}{\underset{|}{C}}-O-\overset{O}{\overset{\|}{C}}-R^1 \\
R^2-\overset{O}{\overset{\|}{C}}-O-\underset{|}{C}-H \\
H-\underset{|}{\overset{|}{C}}-O-\overset{O}{\overset{\|}{C}}-R^3 \\
H
\end{array}
$$

R^1, R^2 and R^3 are long chain hydrocarbons.

a Draw out the structure and circle the ester links.

b The R groups can be saturated or unsaturated.
Explain what this means.

c When a fat or oil is hydrolysed a type of acid is formed.
  i State what is meant by *hydrolysis*.
  ii Name the type of acid formed.
  iii Name the other product formed.

3 The ester molecules found in fats and oils are also known as triglycerides. The structures of three triglycerides and their melting points are shown.

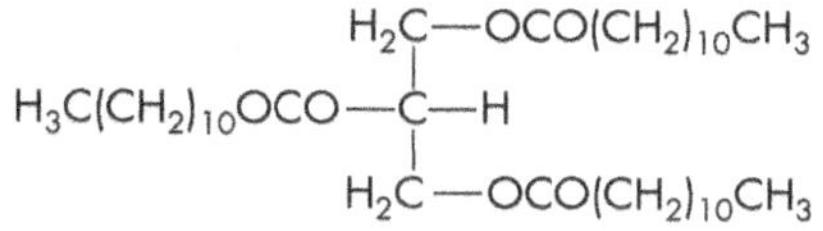

$$H_2C-OCO(CH_2)_{10}CH_3$$
$$H_3C(CH_2)_{10}OCO-C-H$$
$$H_2C-OCO(CH_2)_{10}CH_3$$
triaurin
m.p. 45°C

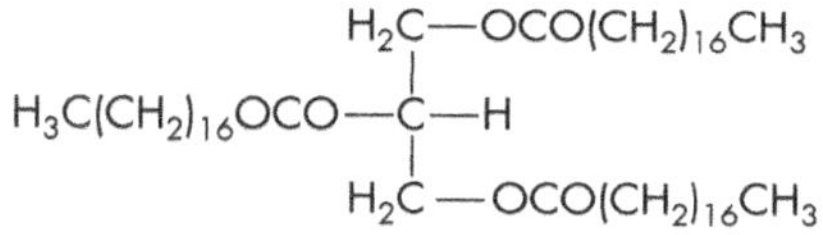

$$H_2C-OCO(CH_2)_{16}CH_3$$
$$H_3C(CH_2)_{16}OCO-C-H$$
$$H_2C-OCO(CH_2)_{16}CH_3$$
tristearin
m.p. 71°C

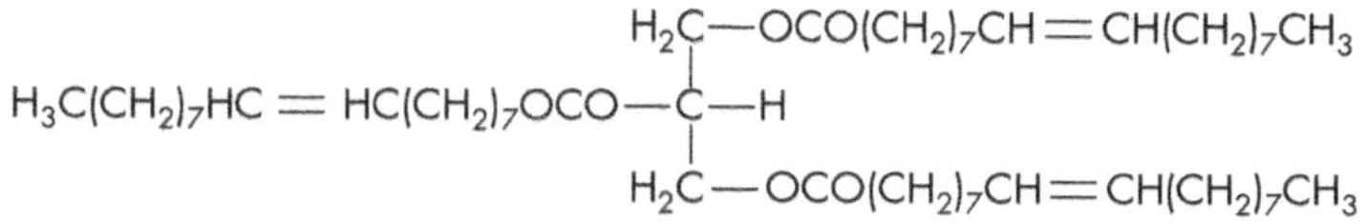

$$H_2C-OCO(CH_2)_7CH=CH(CH_2)_7CH_3$$
$$H_3C(CH_2)_7HC=HC(CH_2)_7OCO-C-H$$
$$H_2C-OCO(CH_2)_7CH=CH(CH_2)_7CH_3$$
triolein
m.p. −4°C

a i Identify the saturated and unsaturated triglycerides.
  ii Explain your answer to part **i**.
b Explain why tristearin has a higher melting point than triaurin.
c Explain why triolein has such a low melting point.
d Fats and oils are mixtures of triglycerides.
  i Suggest which of the three triglycerides fats are likely to contain more of.
  ii Explain your answer to part **i**.
  iii Suggest which of the three triglycerides oils would contain more of.
  iv Explain your answer to part **iii**.

4 The shape of saturated triglyceride molecules can be represented by structure **A** and unsaturated triglycerides by structure **B**.

A

B

a State what is meant by *saturated* and *unsaturated*.

b With reference to the structures, explain why fats have higher melting points than oils.

5 Saturated fatty acids have twice as many hydrogens per molecule as carbons.

The table shows some saturated and unsaturated fatty acids.

Common name	Molecular formula	Saturated/ unsaturated
stearic	$C_{18}H_{36}O_2$	
oleic	$C_{18}H_{34}O_2$	
palmitic	$C_{16}H_{32}O_2$	
palmitoleic	$C_{16}H_{30}O_2$	
linoleic	$C_{18}H_{32}O_2$	

a Copy and complete the table.

b Describe a chemical test you could carry out to determine whether a triglyceride was saturated or unsaturated. Include the result you would obtain.

c The table shows the composition of some common fats and oils we eat.

Fat/oil	Saturated fatty acids (%)	Unsaturated fatty acids (%)	Polyunsaturated fatty acids (%)
lard (pork fat)	40–48	36–51	2–4
butter	33–39	35–40	4–5
olive	6–19	69–85	4–12
soya bean	8–16	21–29	54–67

Medical research has shown that having a greater proportion of polyunsaturated fatty acids in our diet than saturated fatty acids has health benefits.

i Select from the table the fat/oil which is most likely to give the biggest health benefits.

ii Explain your answer to part **i**.

iii Select from the table the fat/oil which is least likely to give health benefits.

iv Explain your answer to part **iii**.

6 Oils can be converted to fats by reacting them with hydrogen.
  a Give the name of this process.
  b Name the type of reaction taking place.
  c The process can be represented as follows.

  $$\text{structure} + 3H_2 \xrightarrow{Ni}$$

  i Draw the structure of the product formed in this reaction.
  ii Suggest what the role of nickel is in the process.
  iii Explain why this process results in a solid being formed.

7 a Outline the importance of fats and oils in our diet.
  b The structures of two vitamins are shown.
  Vitamin A

  Vitamin C

  i Suggest which of the two vitamins is more likely to be soluble in fats and oils.
  ii Explain your answer to part **i**.

7 Soaps, detergents and emulsions

Exercise 7A Soaps and detergents

1 The equation summarises how soap is made.

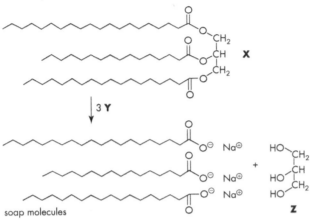

a i State what type of molecule **X** is.
  ii Name reagent **Y**.
  iii Name product **Z**.
b State the type of reaction taking place when reagent **Y** reacts with **X**.
c The term 'soap molecule' is often used.
  Explain why this is not strictly correct.

2 The diagram shows a model of a soap molecule. The tail of the molecule is hydrophobic and the head is hydrophilic.

a i State what is meant by *hydrophobic*.
  ii Explain why the tail of the molecule is hydrophobic.
b i State what is meant by *hydrophilic*.
  ii Explain why the head of the molecule is hydrophilic.
c Explain fully, with the use of diagrams, how soap cleans dirt from your skin.
d One type of soap molecule is called sodium stearate and in soft water it easily forms a lather. Some parts of the country have water which is called hard. This is because it contains magnesium and calcium ions. In these areas a lather is difficult to form and a scum is formed instead.
  i State what type of compound sodium stearate is.
  ii Suggest what the chemical name for scum is.
  iii Suggest whether scum is soluble or insoluble in water.
e State which cleaning agent can be used in hard water areas to avoid the formation of scum.

3 The diagram shows the shape of a detergent molecule.

a Draw out the molecule and label the hydrophobic and hydrophilic parts.
b i Explain why the part of the molecule you labelled hydrophilic acts this way.
  ii Explain why the part of the molecule you labelled hydrophobic acts this way.

4 Sixty percent of detergents in use are anionic detergents. This means the active part of the detergent has a negative charge. Some have no charge and are termed non-ionic. They are used as fabric softeners. Cationic detergents have a positive charge on the active part.

a Identify the following detergent molecules as anionic, cationic or non-ionic.

b Copy structure **C** and circle and label the hydrophilic and hydrophobic parts of the molecule.

c Part of the structure of a non-ionic detergent is shown.

Explain why this part of the molecule is soluble in water despite it not being ionic.

d Explain fully, with the aid of diagrams, how detergent molecules clean dirt from clothes.

e Explain why detergents are particularly useful in hard water areas.

Exercise 7B Emulsions

1 Milk is a natural emulsion of fat and water. A protein called casein acts as an emulsifier.

a i State what is meant by the term *emulsion*.
  ii State the function of an emulsifier.

b The structure of lecithin, found naturally in egg yolks, is shown.

i Lecithin is known as a diglyceride.
   Suggest why this is.
ii Suggest why lecithin can act as an emulsifier.
iii The ingredients for homemade mayonnaise are shown.

Homemade mayonnaise recipe

2 large egg yolks
3 tablespoons lemon juice
$\frac{1}{4}$ teaspoon salt
pinch of white pepper
1 cup oil

Suggest why the oil and lemon juice can combine in mayonnaise but would not normally mix.

2 The structure of a diglyceride of palmitic acid is shown.

$$\begin{array}{l} H-\underset{\underset{H}{|}}{\overset{H}{|}}C-O-\overset{O}{\underset{\|}{C}}-C_{15}H_{31} \\ H-\underset{|}{C}-C-\overset{O}{\underset{\|}{C}}-C_{15}H_{31} \\ H-\underset{\underset{H}{|}}{C}-O-H \end{array}$$

a i Suggest what is meant by the term *diglyceride*.
   ii Copy the structure and identify and label the hydrophilic part of the molecule.
b Explain, with reference to the structure of the diglyceride, why it can act as an emulsifier.

3 The food label for a vegetable fat spread shows it contains a variety of ingredients. The emulsifiers are highlighted.

7 Soaps, detergents and emulsions

a Explain the function of an emulsifier.

b The structural formulae of glycerol and a fatty acid are shown.

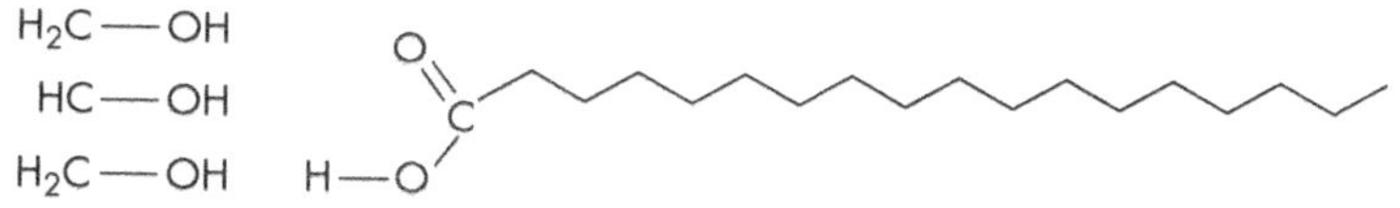

H_2C-OH
$HC-OH$
H_2C-OH

glycerol

fatty acid

i Draw the structural formula of a monoglyceride which could be formed from the structures shown.

ii Draw the structural formula of a diglyceride which could be formed from the structures shown.

c State how mono- and diglyceride molecules differ from fat and oil molecules.

4 The food additive E471 is the most commonly used emulsifier in foods. It contains mono- and diglycerides.

a State the function of an emulsifier.

b The structural formulae of some compounds are shown.

A

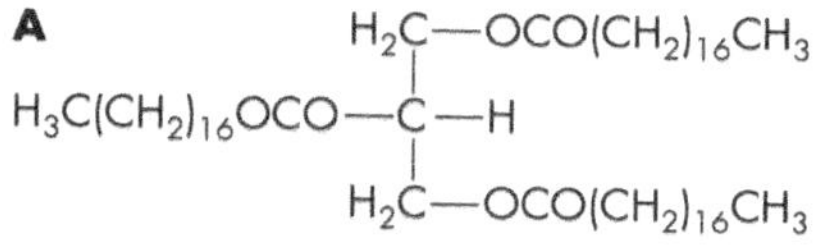

$H_3C(CH_2)_{16}OCO-\overset{H_2C-OCO(CH_2)_{16}CH_3}{\underset{H_2C-OCO(CH_2)_{16}CH_3}{C-H}}$

B

$H_3C-(CH_2)_{16}-C-O-CH$ with $H_2C-O-C-(CH_2)_{16}-CH_3$ and $O-C-(CH_2)_{16}-CH_3$, CH, CH, CH_2HC, OH

C

$HO-\overset{H}{\underset{H}{C}}-H$
$HO-\overset{}{\underset{}{C}}-H$
$H-\overset{}{\underset{H}{C}}-O-\overset{O}{\underset{}{C}}-C_{17}H_{35}$

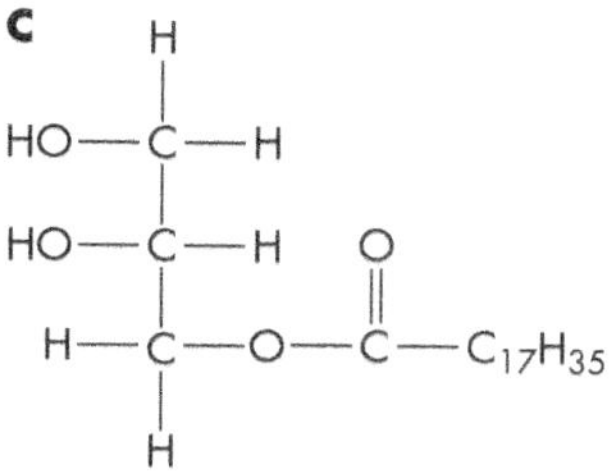

D

$H_3C(CH_2)_7HC=HC(CH_2)_7OCO-\overset{H_2C-OCO(CH_2)_7CH=CH(CH_2)_7CH_3}{\underset{H_2C-OCO(CH_2)_7CH=CH(CH_2)_7CH_3}{C-H}}$

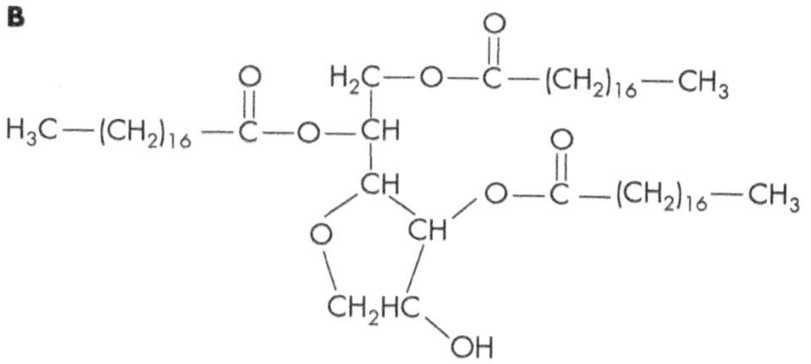

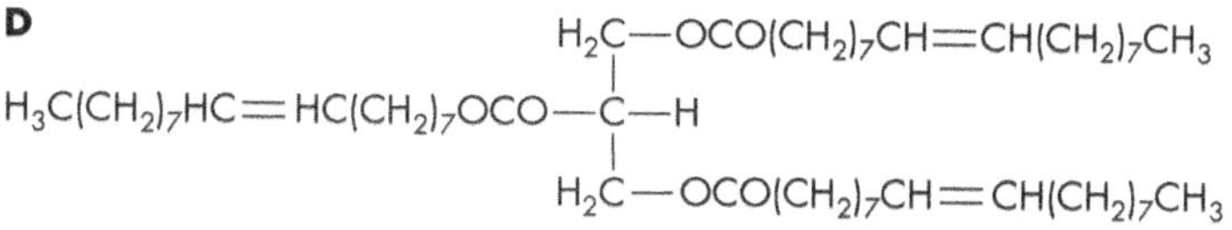

i Select those which could be used as emulsifiers.

ii Explain your choices in part **i**.

7 Soaps, detergents and emulsions

8 Proteins

Exercise 8A Amino acids and proteins

1 Proteins are an essential part of our diet.
State **two** functions of proteins in our bodies.

2 The structure of the amino acid phenylalanine is shown.

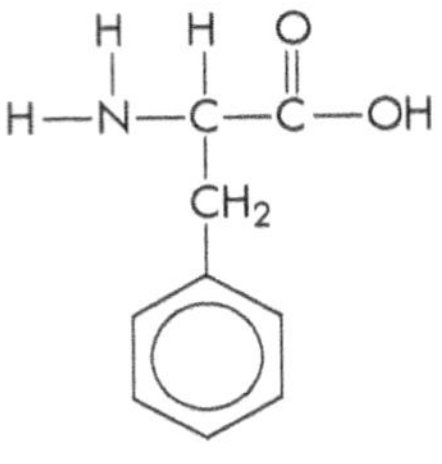

$$\text{H—N(H)—C(H)(CH}_2\text{C}_6\text{H}_5\text{)—C(=O)—OH}$$

a State why our bodies need amino acids.
b Copy the structure and circle the carboxylic acid group in red and the amino group in blue.
c Phenylalanine is known as an essential amino acid.
   i State what is meant by *essential amino acids*.
   ii State how the body obtains essential amino acids.

3 α-amino acids have the amino group attached to the carbon which also has the acid group attached.
Which of the following is an α-amino acid?

A: benzene ring with NH$_2$ and COOH substituents

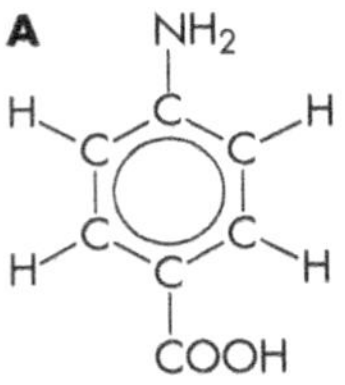

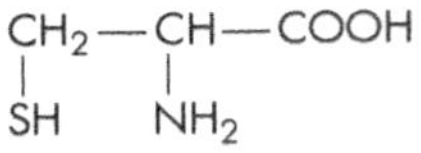

B: $CH_2\text{—}CH\text{—}COOH$ with SH and NH$_2$ substituents

C: $CH_3\text{—}CH\text{—}COOH$ with $CH_2\text{—}NH_2$ substituent

D: benzene ring with NH$_2$ and COOH substituents

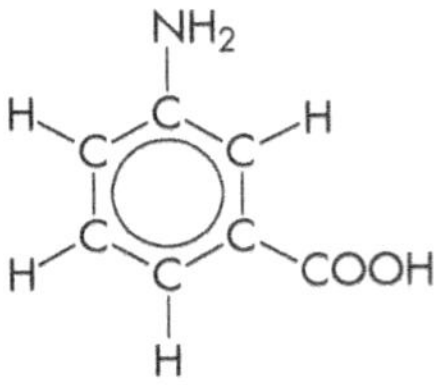

4 Proteins are large molecules formed when many amino acids join.
a Draw out the two amino acids below and show how they join.

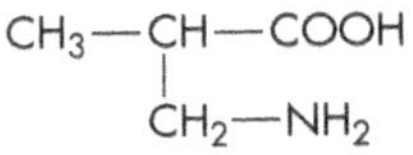

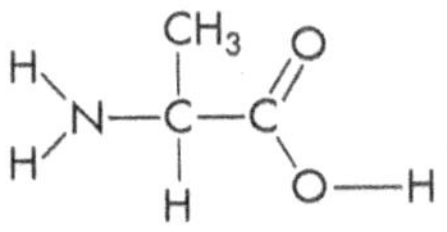

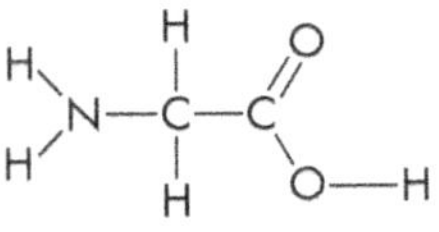

(Two amino acid structures shown: one with CH$_3$ side chain, one with H side chain)

48 8 Proteins

b Amino acids in a protein are joined by an amide link.

   i State the other name by which this link is known.

   ii Circle the link on the structure you drew in part **a**.

c Name the type of reaction taking place when amino acids join.

5 a Which of these is a peptide link?

b Give another name for the peptide link.

6 The diagrams show protein molecules held together by an intermolecular force of attraction, shown by the broken line. Diagram **A** shows a pleated sheet protein while diagram **B** shows a spiral protein.

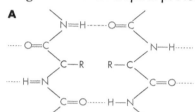   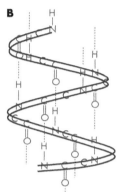

a Name the intermolecular force of attraction shown in the diagrams.

b Hair and skin have protein in their fibres. Skin fibres can be stretched and then return to their normal shape. Hair fibres cannot be stretched like skin.

   i Suggest which type of protein, **A** or **B**, is found extensively in skin.

   ii Explain your answer to part **i**.

   iii Suggest which type of protein, **A** or **B**, is found extensively in hair.

   iv Explain your answer to part **iii**.

7 In egg white the protein is an example of a globular protein in which the protein chains are wrapped around each other and form a ball of molecules.

a Suggest which forces of attraction exist between the protein chains.

b When an egg is heated the protein molecules uncurl and change shape.

   i Suggest what happens to the intermolecular forces between the protein chains when the egg is heated.

   ii Name the process which happens when a protein changes shape in this way.

8 The tough protein which holds meat together is called collagen. The protein chains are wrapped round each other in spirals.

a Suggest what happens to the intermolecular forces between the protein chains.

b Name the process taking place.

8 Proteins

9 Enzymes can break down proteins in the body.
  a i State what type of compound an enzyme is.
     ii What are enzymes acting as in this process?
     iii State the term used for the process of breaking down a protein by enzyme action in the body.
     iv Name the type of reaction taking place.
  b The structure shown is the amino acid residue of a protein chain.

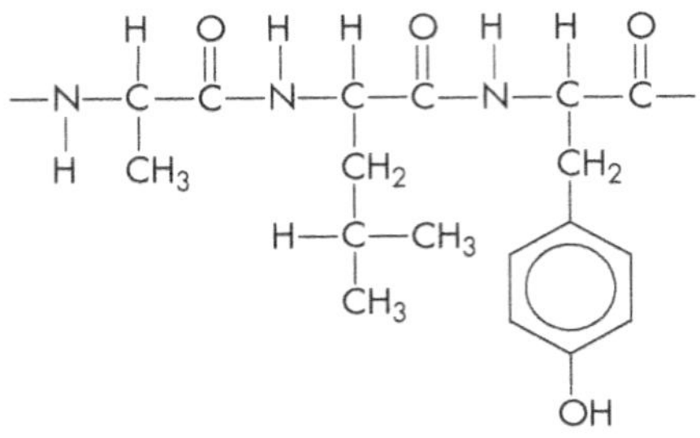

  Draw out the three amino acids in the structure.
  c There are only 20 amino acids.
    Explain why there are thousands of different proteins in your body.

9 Oxidation of foods

Exercise 9A Oxidation of alcohols

1 a State which of the following alcohols cannot be oxidised.

> **Hint** Primary alcohols can be oxidised first to aldehydes then to carboxylic acids.

A: H-C(H)(H)-C(H)(H)-C(H)(H)-O-H

B: H-C(H)(H)-C(H)(O-H)-C(H)(H)-H

C: H-C(H)(H)-C(H)(H)-C(H)(O-H)-C(H)(H)-C(H)(H)-C(H)(H)-H

D: H-C(H)(H)-C(CH3)(O-H)-C(H)(H)-H

b Explain your answer to part **a**.

c For each of the alcohols in part **a** which can be oxidised, name the products.

2 a Name **two** reagents which can be used to oxidise alcohols.

b For each reagent you named in part **a**, describe how you would carry out the oxidation and what you would observe.

3 Different alcohols were added to acidified dichromate solution and warmed. The results are recorded in the table below below.

Alcohol	Obervations
Alcohol A	Colour change from orange to green
Alcohol B	Colour change from orange to green
Alcohol C	No reaction

State what conclusions can be drawn about the type of alcohol added to each test tube.

> **Hint** The oxygen:hydrogen ratio in a compound can tell you if oxidation or reduction has taken place during a reaction. If you're not sure about ratios, divide.
>
> Example: 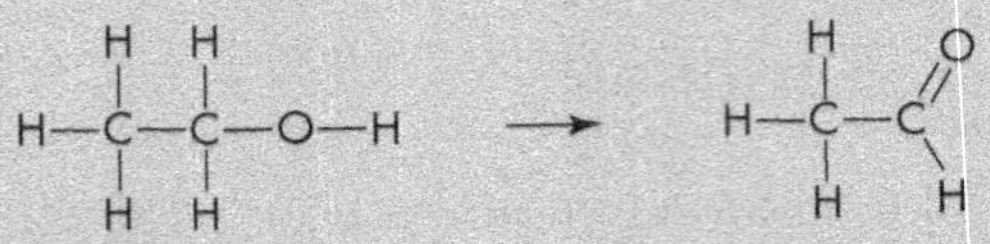
>
> ethanol → ethanal
> Oxygen : hydrogen   1 : 6   1 : 4
>
> An oxygen:hydrogen ratio of 1:4 ($\frac{1}{4}$ = 0·25) is greater than an oxygen:hydrogen ratio of 1:6 ($\frac{1}{6}$ = 0·17). An increase in the oxygen:hydrogen ratio during a reaction means oxidation has taken place.
>
> If the oxygen:hydrogen ratio decreased then a reduction reaction would have taken place.

4 a Use oxygen:hydrogen ratios to prove that each step in the reaction shown below is oxidation.

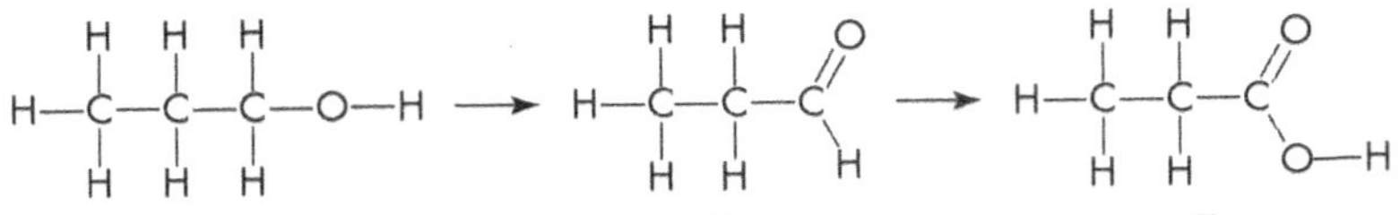

X → Y → Z

b Name **X**, **Y** and **Z**.

5 a Use oxygen:hydrogen ratios to prove that each step in the reaction shown below is reduction.

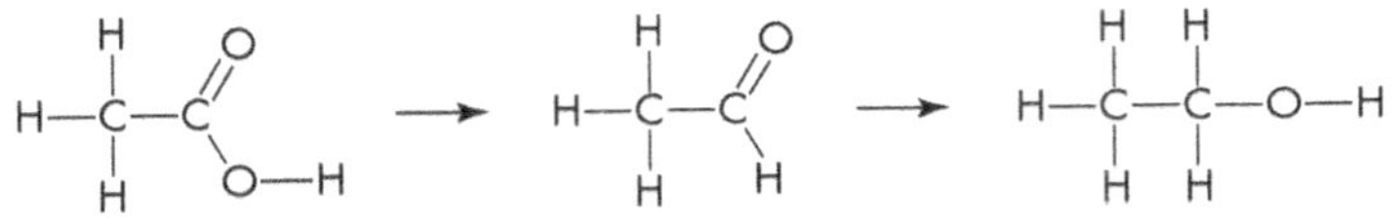

X → Y → Z

b Name **X**, **Y** and **Z**.

6 Edible oils can go rancid if left exposed to the air for too long.

a State what happens to the oil to make it go rancid.

b Vitamin C is commonly added to foods as an antioxidant.
State how an antioxidant works.

c The ion-electron equation showing how vitamin C acts as an antioxidant is shown below.

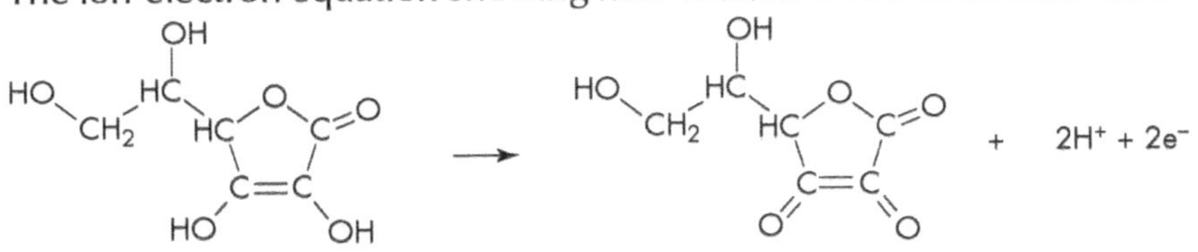

vitamin C → (oxidised form) + $2H^+ + 2e^-$

i Use oxygen:hydrogen ratios to prove that vitamin C is acting as an antioxidant.

ii State what else in the ion-electron equation indicates that vitamin C is acting as an antioxidant.

Exercise 9B Naming aldehydes and ketones

> **Hint** The rules for naming aldehydes, ketones, alcohols and carboxylic acids are similar.
>
> The name ending will be different but the numbering and naming of branches and the position of the functional groups are indicated in the same way.

Example
Name the compound with the following structure.

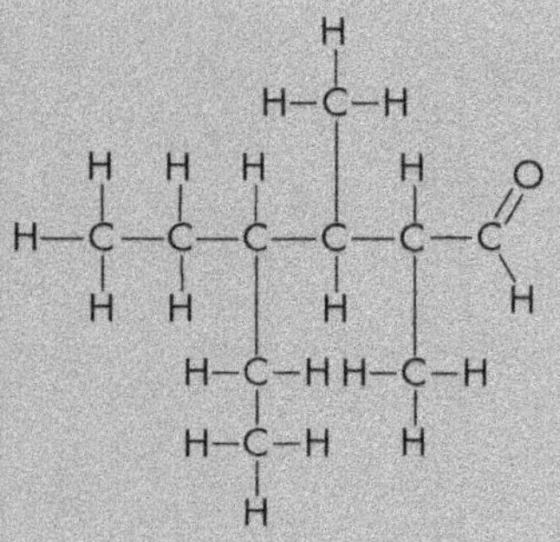

- The compound has the aldehyde functional group so the name will end in **-al**.
- The longest chain has six carbon atoms – the main part of the name is **hexanal**.
- Branches are numbered by counting the carbons from the functional group end.
- There is one **ethyl** branch on the fourth carbon (**4-ethyl-**), and two **methyl** branches, one on each of the second and third carbon atoms (**2,3-dimethyl**).
- The branches are arranged in alphabetical order in the name.

  Name: 4-ethyl-2,3-dimethylhexanal
- When naming ketones and alcohols, the position of the functional group is indicated in the middle of the name of the longest straight chain, e.g. 3-methylhexan-2-one.

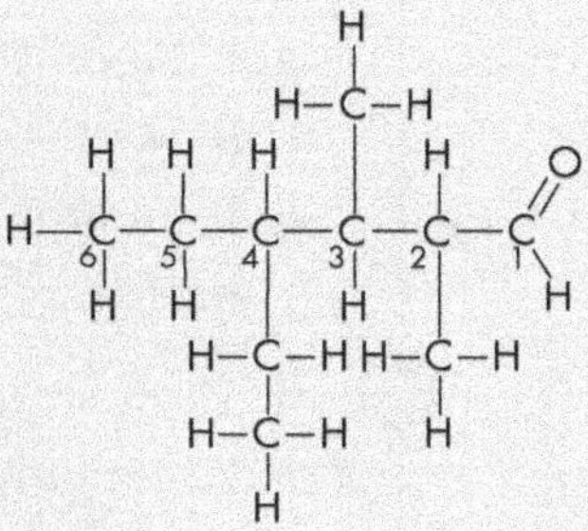

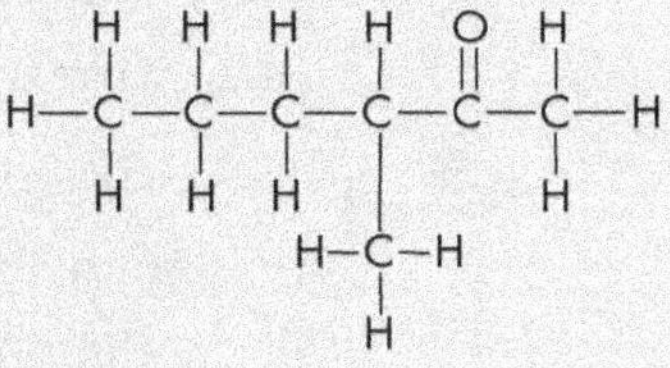

1 Aldehydes and ketones can be found in many foods.

  a State the characteristic which some aldehydes and ketones give food.

  b i Name the functional group present in all aldehydes and ketones.

  ii Draw the functional group named in part **i**.

  c Explain why, despite having the same functional group, aldehydes and ketones have different properties.

2 The following is a list of aldehydes and ketones.

  methanal   propanone   benzaldehyde   ionone   ethanal   carvone

  a Identify the aldehydes in the list above.

  b Identify the ketones in the list above.

  c i Write the molecular formula for ethanal.

  ii Draw the full structural formula for ethanal.

  iii Write the shortened structural formula for ethanal.

9 Oxidation of foods

d i Write the molecular formula for propanone.
  ii Draw the full structural formula for propanone.
  iii Write the shortened structural formula for propanone.

3 The structures below show aldehydes and ketones.

A, B, C, D [structures of aldehydes and ketones shown]

a i Identify the aldehydes.
  ii Name the aldehydes you identified in part **i**.
b i Identify the ketones.
  ii Name the ketones you identified in part **i**.

4 a Draw structural formulae for the following compounds.
  i 4-methylpentanal   **ii** 3-methylbutanal
  iii 2,3-dimethylhexanal   **iv** 3-ethyl-4-methylheptanal
 b Write molecular formulae for each of the compounds in part **a**.
 c State what type of compound each of the examples in part **a** are.

5 a Draw structural formulae for the following compounds.
  i 3-methylbutan-2-one   **ii** 4-ethylhexan-3-one
  iii 2,2-dimethylpentan-3-one   **iv** 5-ethyl-2-methyloctan-4-one
 b Write molecular formulae for each of the compounds in part **a**.
 c State what type of compound each of the examples in part **a** are.

6 Look at the compounds in Questions 4 and 5.
Identify the pairs of isomers.

7 a Name the family of compounds to which pentanone belongs.
 b Two straight chain isomers can be drawn for pentanone.
   i Draw the two straight chain isomers of pentanone.
   ii Name each of the isomers you drew in part **i**.
 c A branched chain isomer can be drawn for pentanone.
   i Draw the branched chain isomer for pentanone.
   ii Name the isomer you drew in part **i**.

d i Draw the structure of the straight chain aldehyde which is an isomer of pentanone.

　ii Name the isomer you drew in part **i**.

e Three branched chain isomers can be drawn for the aldehyde with the same molecular formula as pentanone.

　i Draw the three branched chain isomers.

　ii Name each of the isomers you drew in part **i**.

Exercise 9C Reactions and properties of aldehydes and ketones

1 a Select from the following list of compounds those which can be oxidised.

　i Ethanal　　**ii** Propanal　　**iii** Propanone

　iv 2-methylpropanone　　**v** Butanal

b Name the family to which the compounds you identified in part **a** belong.

c Name the products formed when each of the compounds you selected in part **a** are oxidised.

d Name **three** reagents which can be used to oxidise the compounds you selected in part **a**.

e Describe how you would carry out the oxidation reaction using each of the reagents you gave in part **d**. Include the colour change you would see in each case.

2 Part of the ion-electron equation for methanal forming methanoic acid is shown.

HCHO(ℓ) + _____ → HCOOH + _____ + _____

a i Copy and complete the ion-electron equation.

　ii State what type of reaction is taking place.

b The reaction can be carried out using acidified dichromate.

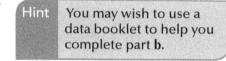

Hint You may wish to use a data booklet to help you complete part **b**.

　i Write the ion-electron equation for the dichromate ion ($Cr_2O_7^{2-}$) forming the chromium 3+ ion (Cr^{3+}) when it reacts with methanal.

　ii State what type of reaction is happening in part **i**.

3 The photos show aldehydes being oxidised by various reagents. The left-hand test tube in each case is the reagent on its own and the right-hand test tube shows what happens when the aldehyde is added and the mixture warmed.

a

The mixture has changed from orange to green.

　i Name the reagent.

　ii Write the ion-electron reduction equation to show what is happening to the reagent.

b

The mixture has changed from blue to brown.

   i Name the reagent.

   ii Write the ion-electron reduction equation to show what is happening to the reagent.

c

The mixture has changed from colourless to silver.

   i Name the reagent.

   ii Write the ion-electron reduction equation to show what is happening to the reagent.

4 Two colourless liquids are thought to be an aldehyde and a ketone.

Describe a chemical test which could be carried out on the liquids to tell them apart.

Include:
- the name of the reagent
- how the test is carried out
- what you would see happening with the aldehyde
- what you would see happening with the ketone.

5 The structural formulae and boiling points of some compounds are shown.

propan-1-ol
b.p.: 97°C

propanone
b.p.: 56°C

propanal
b.p.: 49°C

  a Explain why there is a big difference in the boiling points of propanone and propanal compared with propan-1-ol, despite the similarities in their molecular mass.

  b **i** State how the volatility of the compounds compare.

     ii Explain why volatility of compounds is important in cooking.

  c **i** Explain why aldehydes and ketones are soluble in water.

     ii Explain why the solubility decreases as the molecules increase in size.

Exercise 9D Flavour molecules

1 The structural formulae of some flavour molecules are shown. Their structures have some common features but also groups which give them different properties.

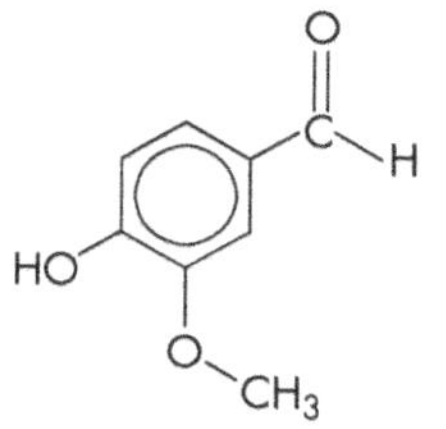

vanillin: found in vanilla pods

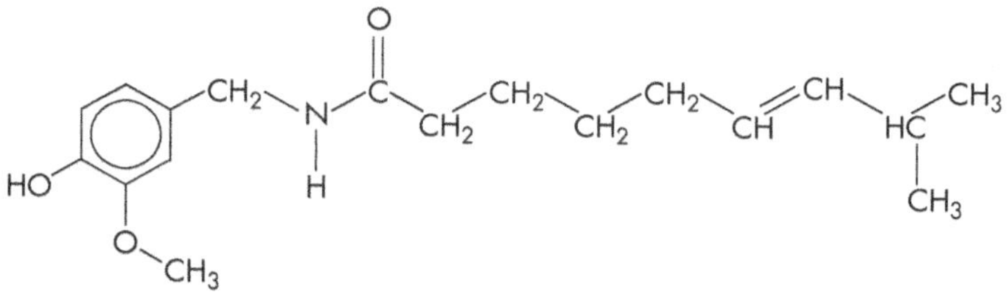

capsaican: one of the molecules which gives chilli its hot taste

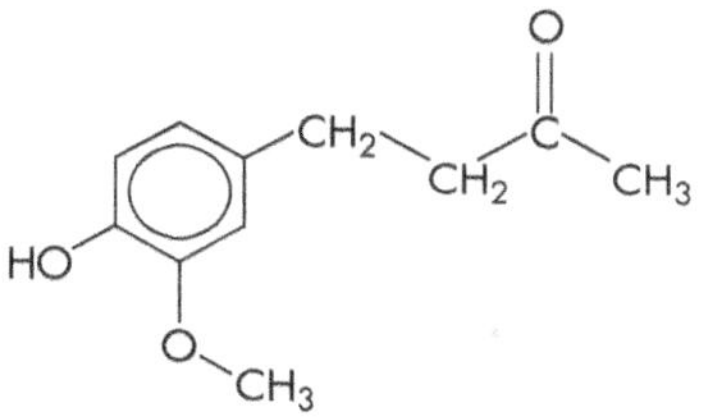

zingerone: found in ginger root

- **a i** Identify the ketones.
  - **ii** Identify the aldehyde.
- **b i** Identify the compound which is most likely to be soluble in water.
  - **ii** Explain your answer to part **i**.
- **c i** Identify the compound which is most likely to be soluble in oil.
  - **ii** Explain your answer to part **i**.
- **d** Suggest why swilling your mouth with water does little to relieve the discomfort of eating food with too much chilli.

2 The structural formulae of two flavour molecules, limonene and furaneol, are shown.

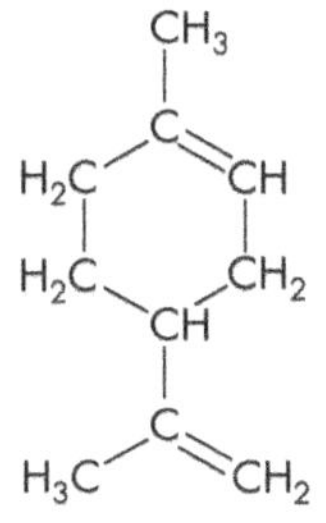

limonene

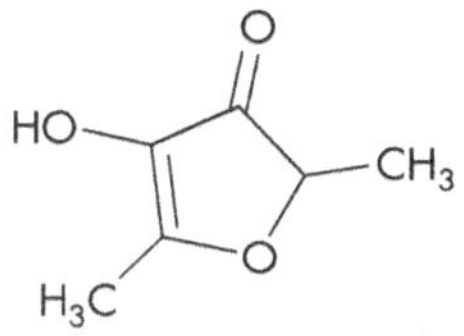

furaneol

- **a** Explain why furaneol is soluble in water but limonene is not.
- **b** Explain why limonene is soluble in hydrocarbons such as hexane.
- **c** Suggest why furaneol has some solubility in oil-based solvents.

3 Citronellol and undecanal are commonly used in perfumery. Their structures are shown below.

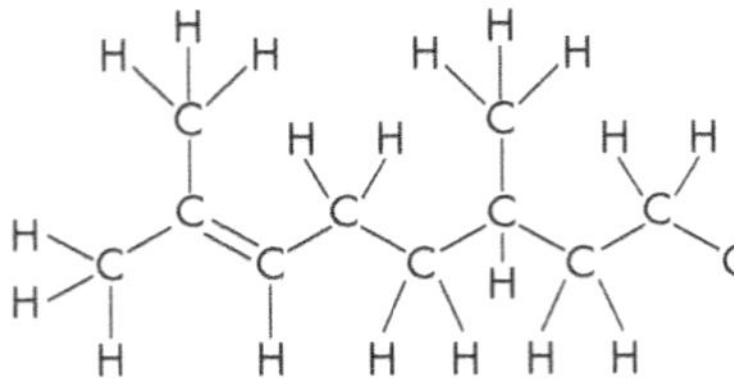

citronellol

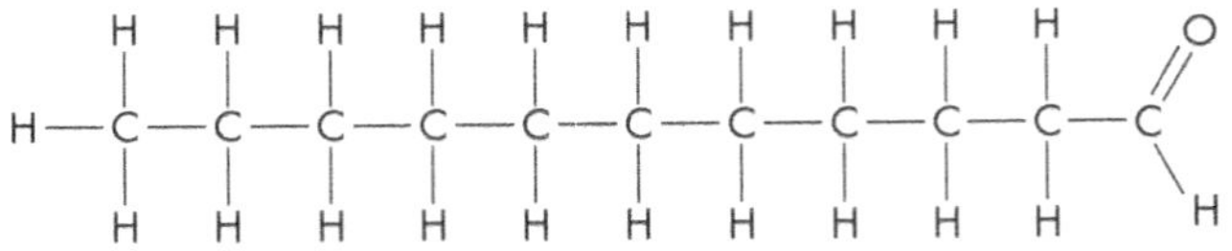

undecanal

a i State what type of compound undecanal is.

ii State what type of compound citronellol is.

b Explain why citronellol is much less volatile than undecanal.

4 a A cookery book suggests that in order to preserve its flavour asparagus should be cooked by coating it in oil and grilling it rather than by boiling it in water.

State what this suggests about the solubility of the flavour molecules in asparagus.

b The cookery book also recommends boiling broccoli in water rather than cooking it in oil.

State what this suggests about the flavour molecules in broccoli.

10 Fragrances

Exercise 10A Essential oils

 Essential oils can be used to flavour foods and give cleaning products a pleasant smell.

   a **i** State what an essential oil is.

   ii Give **two** other uses of essential oils.

   b The diagram shows how essential oils can be extracted from orange peel in the laboratory. The orange peel is placed in the wire gauze basket. The essential oil and water are collected in the test tube in the ice bath.

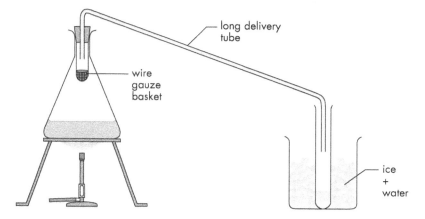

   i State what this method of extraction tells us about the volatility of essential oils.

   ii Explain your answer to part **i**.

   iii Describe what you would see in the test tube in the ice bath.

   iv State what your observation in part **iii** tells you about the solubility of essential oils in water.

   > **Area 4** – Researching Chemistry: Common chemical apparatus
   >
   > **c** **i** Name the glass container being heated.
   >
   > **ii** State what the delivery tube is acting as.

2 Essential oils contain a mixture of terpenes.

a The structure of the compound which terpenes are made up from is shown.

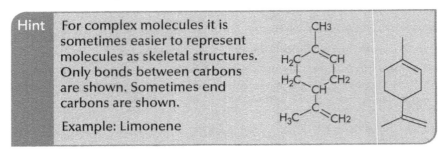

Name the compound.

> **Hint** For complex molecules it is sometimes easier to represent molecules as skeletal structures. Only bonds between carbons are shown. Sometimes end carbons are shown.
>
> Example: Limonene

b i Terpenes can react inside plants to form alcohols, aldehydes and ketones, often collectively known as terpenoids.

Identify the following terpene skeletal structures as hydrocarbons, alcohols, aldehydes or ketones.

A

B

C

D

E

ii Name the type of reaction taking place when terpenoids are formed.

iii Suggest why it is recommended that spices be kept whole and only ground up or grated when needed for cooking.

3 The table shows the different classes of terpene.

Class of terpene	Number of isoprene units	Formula
monoterpenes	2	$C_{10}H_{16}$
sesquiterpenes	3	$C_{15}H_{24}$
diterpenes	4	$C_{20}H_{32}$
sesterterpenes	5	$C_{25}H_{40}$
triterpenes	6	$C_{30}H_{48}$
tetraterpenes	8	$C_{40}H_{64}$
polyterpenes	many	$(C_5H_8)n$

a Identify the class of terpene to which each of the following belongs.

b Copy out structure **A** and circle the isoprene units.

c i Geraniol is a monoterpene.

   Identify which of the structures **A–G** is geraniol.

 ii Squalene is a triterpene.

   Identify which of the structures **A–G** is squalene.

4 The structures of three terpenes are shown.

a Identify: **i** limonene   **ii** carveol   **iii** carvone

b Name the functional groups in **A** and **B** which identifies them.

c The three terpenes have similar structures but limonene smells of oranges and carveol and carvone smell of spearmint.

Suggest what causes this difference in smell.

11 Skin care

Exercise 11A Free radical chain reactions

1 Premature ageing of the skin, also known as photoageing, is due to exposure to the sun's rays.

   a State **one** other effect exposure to the sun can have on the skin.

   b **i** Name the type of radiation which causes damage to the skin.

      ii State what feature the radiation in part **i** has which causes damage to the skin.

   c **i** State what effect the sun's radiation can have on the molecules which make up our skin.

      ii Sun protectors can be applied to the skin to reduce the amount of damage caused by exposure to the sun.

        State how these lotions protect the skin.

2 The diagram shows an experiment which can be carried out to show how ultraviolet (UV) light can cause a chemical reaction to occur.

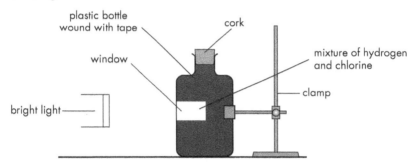

   a **i** State why no reaction occurs when only the laboratory lights are on but an explosive reaction occurs when the bright light is switched on.

      ii The reaction between chlorine and hydrogen is known as a free radical chain reaction.

        State what is meant by a *free radical*.

   b The mechanism for the reaction is outlined below.

Reaction	Step
$Cl_2(g) \rightarrow 2X$	1
$X + H_2(g) \rightarrow HCl(g) + Y$	2
$Y + Cl_2(g) \rightarrow HCl(g) + X$	2
$X + Y \rightarrow HCl(g)$	3

     i Write formulae for **X** and **Y**.

     ii Name steps 1, 2 and 3.

3 Bromine solution reacts slowly with hexane but if left in the sunlight the reaction takes place much more quickly.

   a i Suggest why the reaction is quicker in sunlight.

     ii State what you would see happening to indicate that the bromine and hexane are reacting.

   b In the first step in the reaction, bromine molecules form bromine radicals.

     i Write an equation for this reaction step.

     ii State the name given to this step in the reaction.

   c The equations show what happens in the second step in the reaction.

$$C_6H_{14} + X \rightarrow C_6H_{13}Br + Y$$
$$Y + C_6H_{13}Br \rightarrow C_6H_{14} + X$$

     i Write formulae for X and Y.

     ii State the name given to this step in the reaction.

   d The equation shows a step in the reaction which stops the whole reaction.

$$H\cdot + Br\cdot \rightarrow HBr$$

     State the name given to this step in the reaction.

   e State the name given to this type of reaction.

4 Ozone (O_3) in the stratosphere protects us from some of the harmful effects of the sun's rays. When chlorine-containing compounds reach the stratosphere they break down and react with ozone.

Some of the steps in the reaction sequence are shown, but not in the order they would happen.

A $Cl\cdot + O_3 \rightarrow ClO\cdot + O_2$

B $O\cdot + O_3 \rightarrow 2O_2$

C $CF_3Cl \rightarrow CF_3 + Cl\cdot$

D $ClO\cdot + O\cdot \rightarrow Cl\cdot + O_2$

   a Put the steps in the order they occur.

   b Identify:

     i the initiation step

     ii the propagation steps

     iii the termination step.

   c State the name given to this type of reaction.

5 Many cosmetics contain free radical scavengers.

   a i State what is meant by a *free radical scavenger*.

     ii Suggest how the addition of free radical scavengers to cosmetics could have a positive effect on the skin.

     iii Suggest another name for a free radical scavenger.

   b Name **two** other products which contain free radical scavengers.

12 Getting the most from reactants

Exercise 12A The chemical industry and green chemistry

1 Copy and complete the sentence.

Industrial processes are designed to maximise _____ and minimise the impact on the _____.

2 The flow diagram below shows a typical traditional industrial process.

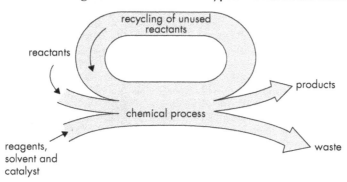

a Suggest what could be done to reduce waste and at the same time save money.

b An environmental factor, known as the E factor, can be used to measure the efficiency of industrial processes.

$$\text{E factor} = \frac{\text{mass of waste}}{\text{mass of products}}$$

Oil refining has an E factor of almost zero, while the typical E factor in the pharmaceutical industry is between 25 and 100.

  i Suggest which of oil refining or the pharmaceutical industry is the most waste-efficient.

  ii Explain your answer to part **i**.

c The equation shows the production of carbonyl chloride ($COCl_2$) used to make a polycarbonate.

$$CO(g) + Cl_2(g) \longrightarrow \underset{\text{carbonyl chloride}}{\overset{Cl}{\underset{Cl}{\diagup}}}C=O \ (g)$$

  i Suggest why this process is not considered to be environmentally friendly.

  ii An alternative route to the polycarbonate is by using dimethyl carbonate.

$$2CH_3OH(\ell) + \tfrac{1}{2}O_2(g) + CO(g) \xrightarrow{CuCl_2} \underset{\text{dimethyl carbonate}}{\overset{CH_3O}{\underset{CH_3O}{\diagup}}}C=O(\ell) + H_2O(\ell)$$

Suggest why this is considered a more environmentally friendly process.

3 Rare earth elements are key components of everyday items like mobile phones and laptops and are also essential for making magnets used in wind turbines and batteries used in electric cars.

The map shows where the main rare earth deposits in the world are mined.

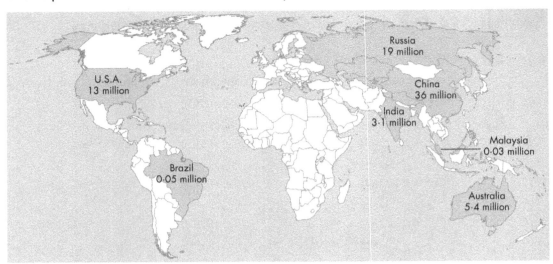

   a Suggest why there is concern in the European Union about the:

   i availability of rare earth metals

   ii sustainability of rare earth metals

   iii cost of rare earth metals

   b Suggest why countries like India are exporting less of their rare earth metals.

   c Suggest why the UK Government is keen to support companies that are exploring the sea bed for rare earth metal deposits.

4 **a** In industry, reactions can be in the gas phase or in solution with various solvents.

   Suggest why gas phase reactions are often preferred to reactants in solution.

   b Suggest why farmers prefer pesticides which break down in sunlight in 2–3 days rather than those that last for months.

   c Suggest why the use of biodegradable detergents in the home is preferred by operators of sewage works.

   d Suggest why liquid carbon dioxide is preferred to tetrachloroethene (C_2Cl_4) as a dry-cleaning solvent.

   e Suggest why bioethanol is considered to be a better source of ethene than the oil industry in the future.

Exercise 12B Percentage yield and the atom economy

> **Hint** The relationships used to calculate percentage yield and atom economy can be found in the SQA data booklet.
>
> $$\text{Percentage yield} = \frac{\text{actual yield}}{\text{theoretical yield}} \times 100$$
>
> $$\text{Atom economy} = \frac{\text{mass of desired product}}{\text{total mass of reactants}} \times 100$$

1. The reaction sequence below summarises two routes to ibuprofen, a medicine which reduces inflammation and gives pain relief.

   a Suggest why the BHC route is the preferred industrial route for making ibuprofen.

   b Compare the likely energy consumption and waste production of the two routes.

   c The overall percentage yield of a reaction with a number of steps can be calculated.

   For example, if each step in a reaction gave a 90% yield, then:

   A one-step reaction would give a 90% yield.

   A two-step reaction would give a $100 \times (0.9 \times 0.9) = 81\%$ yield.

   A three-step reaction would give a $100 \times (0.9 \times 0.9 \times 0.9) = 73\%$ yield.

   Etc.

   i   If each step in the Boots route and the BHC route gave a 75% yield, calculate the overall percentage yield for both routes.

   ii Using your answer to part **i**, write a statement linking the number of steps in a reaction to the overall percentage yield.

12 Getting the most from reactants

2 Ethanoic acid can be made industrially by two routes.

Route 1: From naphtha

naphtha + air → ethanoic acid + propanone + methanoic acid + propanoic acid

Route 2: From methanol

methanol + carbon monoxide → ethanoic acid

a Route 1 gives a number of by-products.
  i Suggest **one** benefit from producing these by-products.
  ii Suggest **two** negative aspects regarding the production of by-products.
  iii Suggest **one** reason that route 2 is considered to have a potentially negative impact on the environment.

b Route 2 gives a higher percentage yield of ethanoic acid but the atom economy of the two routes are similar.
  i Explain what is meant by *percentage yield*.
  ii Explain what is meant by *atom economy*.
  iii Explain why the atom economy of both routes can be considered to be similar.

3 Phenol (C_6H_5OH) can be manufactured from benzene (C_6H_6) by two routes.

Route 1:

$C_6H_6 + H_2SO_4 + 2NaOH \rightarrow C_6H_5OH + Na_2SO_3 + 2H_2O$
                                                         sodium sulfite

Route 2:

$C_6H_6 + C_3H_6 + O_2 \rightarrow C_6H_5OH + CH_3COCH_3$
                                                propanone

a i The theoretical yield of phenol by route 1 is 94 g from 78 g of benzene. The actual yield was found to be 73 g.
    Calculate the percentage yield of phenol by route 1.
  ii Calculate the atom economy for the production of phenol by route 1.

b The atom economy for the production of phenol by route 2 is regarded to be 100%.
  Suggest why this is the case for route 2 and not route 1.

4 Calculate the atom economy for the following.

a The production of ethylene oxide (C_2H_4O) from ethene (C_2H_4) by the chlorohydrin process.

$C_2H_4 + Cl_2 + Ca(OH)_2 \rightarrow C_2H_4O + CaCl_2 + H_2O$

b The production of ethanol (C_2H_5OH) by fermentation of glucose ($C_6H_{12}O_6$).

$C_6H_{12}O_6 \rightarrow 2C_2H_5OH + 2CO_2$

Exercise 12C Calculating quantities from balanced equations

> **Hint**
> The concept of the mole allows us to carry out calculations including quantities reacting and being produced.
>
> 1 mol = gram formula mass (GFM)
>
> $\text{mol} = \dfrac{\text{mass}}{\text{GFM}}$
>
> mass = mol × GFM
>
> (The relationship linking moles, mass and GFM is given in the SQA data booklet as $n = m/\text{GFM}$, where n = moles, m = mass and GFM = gram formula mass.)

Example

The ingredients label on a 30 g packet of crisps states that it contains 0·12 g of sodium.

- Step 1: Calculate the number of moles of sodium ions – this will be the same as the number of chloride ions in NaCl (1:1 ratio). This is also the number of moles of NaCl ionic units.

Number of moles of sodium ions = mass/GFM = $\dfrac{0.12}{23}$ = 0·0052 mol

- Step 2: Calculate the mass of NaCl from the number of moles.

mass = mol × GFM   GFM = 23 + 35·5 = 58·5 g

= 0·0052 × 58·5

= 0·30 g

The mass of sodium chloride in a 30 g packet of crisps is 0·30 g.

1 The ingredients label on a packet of breakfast cereal states it contains 0·32 g of sodium in every 100 g. The sodium is in the form of sodium chloride (NaCl).

Calculate the mass of sodium chloride in 100 g of cereal.

2 A packet of low-sodium salt contains 128 g of potassium chloride (KCl).

Calculate the mass of potassium in the packet.

Example

Calculate the mass of water produced when 30·4 g of methane is burned completely in excess oxygen.

Balanced equation: $CH_4 + 2O_2 \rightarrow CO_2 + 2H_2O$

- Step 1: Identify the moles reacting/produced.

| | 1 mol | 2 mol | → | 1 mol | 2 mol |

Key relationship: 1 mol   →   2 mol

- Step 2: Convert to mol (mol = mass/GFM).

$\dfrac{30.4}{16}$ = 1·9 mol   →   3·8 mol

12 Getting the most from reactants

- Step 3: Convert to mass (mass = mol × GFM).

  mass = 3·8 × 18

  = 68·4 g

  Mass of water produced is 68·4 g.

3 Calculate the mass of carbon dioxide produced when 120 g of butane (C_4H_{10}) is burned completely in excess oxygen.

$$C_4H_{10} + \frac{13}{2}O_2 \rightarrow 4CO_2 + 5H_2O$$

4 Calculate the mass of iron produced when 500 kg of iron(II) oxide reacts completely with carbon monoxide.

$$Fe_2O_3 + 3CO \rightarrow 2Fe + 3CO_2$$

5 Calculate the mass of copper metal which would be needed to react with excess silver(I) nitrate solution to produce 3·2 g of silver.

$$Cu + 2AgNO_3 \rightarrow 2Ag + Cu(NO_3)_2$$

Exercise 12D Calculating percentage yield

> **Hint** The relationship used to calculate percentage yield can be found in the SQA data booklet.
>
> $$\text{Percentage yield} = \frac{\text{actual yield}}{\text{theoretical yield}} \times 100$$
>
> The theoretical yield is the yield calculated from the balanced equation.
>
> The actual yield is the quantity of the desired product achieved when the reaction takes place.
>
> In Questions 3–5 in Exercise 12C it is the theoretical yield which is being calculated.

Example

Ethene can be converted into ethanol by passing a mixture of steam and ethene over a catalyst.

$$C_2H_4 + H_2O \rightarrow C_2H_5OH$$

Calculate the percentage yield of ethanol if 57·4 g of ethene produces 72·8 g of ethanol.

- Step 1: Calculate the theoretical yield.

  Identify the moles reacting/produced.

	C_2H_4	+	H_2O	→	C_2H_5OH
	1 mol		1 mol	→	1 mol

  Key relationship: 1 mol → 1 mol

  Convert to mol (mol = mass/GFM). $\frac{57.4}{28} = 2.05$ mol → 2·05 mol

  Convert to mass (mass = mol × GFM).

  Mass = 2·05 × 46

  Theoretical yield = 94·3 g

- Step 2: Calculate the percentage yield

$$\text{Percentage yield} = \frac{\text{actual yield}}{\text{theoretical yield}} \times 100$$

$$= \frac{72 \cdot 8}{94 \cdot 3} \times 100$$

Percentage yield = 77·2%

1 Methyl ethanoate is produced when methanol reacts with ethanoic acid.

$CH_3OH + CH_3COOH \rightleftharpoons CH_3OOCCH_3 + H_2O$

Calculate the percentage yield when 7·9 g of methyl ethanoate was produced from 9·6 g of ethanoic acid.

2 Calculate the percentage yield of ammonia if 850 kg of ammonia is produced when 790 kg of hydrogen reacts with excess nitrogen.

$N_2 + 3H_2 \rightleftharpoons 2NH_3$

3 Aspirin ($C_9H_8O_4$) can be made by reacting salicylic acid ($C_7H_6O_3$) with ethanoic anhydride ($C_4H_6O_3$).

$C_7H_6O_3 + C_4H_6O_3 \rightarrow C_9H_8O_4 + C_2H_4O_2$

Calculate the percentage yield of aspirin if 24·8 g of aspirin is produced by reacting 56·2 g of salicylic acid with excess ethanoic anhydride.

Exercise 12E Calculations involving moles, mass, concentration and volume

> **Hint** The SQA data booklet gives the relationship between:
> - moles (n), concentration (c) and volume of solution (V) as: $n = c \times V$.
>   From this, $c = n/V$ and $V = n/c$.
> - moles (n), mass (m) and gram formula mass (GFM) as: $n = m/\text{GFM}$.
>   From this, $m = n \times \text{GFM}$.
>
> The number of moles reacting can be calculated using balanced equations.

Example

Calculate the mass of potassium hydroxide needed to completely neutralise 50 cm³ of 0·2 mol l⁻¹ sulfuric acid.

$2KOH + H_2SO_4 \rightarrow K_2SO_4 + 2H_2O$

Balanced equation: $2KOH + H_2SO_4 \rightarrow K_2SO_4 + 2H_2O$

- Step 1: Identify the key mole relationship.

  2 mol   1 mol

- Step 2: Calculate the number of moles of H_2SO_4 from the information in the question.

  $n = c \times V$
  $= 0.2 \times 0.05$
  $n = 0.01$ mol

- Step 3: Work out the number of moles of KOH from the mole ratio in step 1, i.e. 2:1.

  0.01 mol of H_2SO_4 available so 0.02 mol of KOH needed.

- Step 4: Convert moles to mass.

  $m = n \times GFM$
  $= 0.02 \times 56.1$
  $= 1.12$ g

So the mass of potassium hydroxide required is 1.12 g.

1 Calculate the mass of magnesium which would be needed to react completely with 250 cm³ of 0.1 mol l⁻¹ sulfuric acid.

$Mg + H_2SO_4 \rightarrow MgSO_4 + H_2$

2 Calculate the mass of calcium carbonate required to react completely with 100 cm³ of 0.05 mol l⁻¹ hydrochloric acid.

$CaCO_3 + 2HCl \rightarrow CaCl_2 + CO_2 + H_2O$

3 Calculate the mass of sodium hydroxide which would be needed to react completely with 40 cm³ of 0.2 mol l⁻¹ sulfuric acid.

$2NaOH + H_2SO_4 \rightarrow Na_2SO_4 + 2H_2O$

> **Hint** In a reaction, given the volume of one solution, its concentration can be calculated if the volume and concentration of the other solution are known.
>
> The relationship used is:
>
> $$\frac{c_1 V_1}{n_1} = \frac{c_2 V_2}{n_2}$$
>
> where c_1 = concentration of reactant 1
>
> c_2 = concentration of reactant 2
>
> V_1 = volume of reactant 1
>
> V_2 = volume of reactant 2
>
> n_1 = number of moles of reactant 1 from balanced equation
>
> n_2 = number of moles of reactant 2 from balanced equation
>
> The relationship can be found in the SQA data booklet.

Example

20.0 cm³ of potassium hydroxide solution was neutralised by 12.0 cm³ of 0.1 mol l⁻¹ sulfuric acid. The balanced equation for the reaction is:

$2KOH + H_2SO_4 \rightarrow K_2SO_4 + 2H_2O$

Calculate the **concentration** of the potassium hydroxide solution.

$$\frac{c_1 V_1}{n_1} = \frac{c_2 V_2}{n_2}$$

So,

$$\frac{c_1 \times 20.0}{2} = \frac{0.1 \times 12.0}{1}$$

So,

$$c_1 = \frac{0.1 \times 12.0 \times 2}{20 \times 1}$$

$c_1 = 0.12 \text{ mol l}^{-1}$

4 Calculate the concentration of a potassium hydroxide solution if 20.0 cm³ of the solution is completely neutralised by 16.0 cm³ of 0.1 mol l⁻¹ sulfuric acid.

$2NaOH + H_2SO_4 \rightarrow Na_2SO_4 + 2H_2O$

5 10.0 cm³ of nitric acid was neutralised by 15.2 cm³ 0.2 mol l⁻¹ sodium hydroxide. Calculate the concentration of the nitric acid.

$NaOH + HNO_3 \rightarrow NaNO_3 + H_2O$

Exercise 12F Reactants in excess

Hint In a reaction, seldom do all of the reactants get used up. One reactant is usually the **limiting reactant** – one which is completely used up. Other reactants which are not completely used up are in **excess**. Calculations must be based on the quantity of limiting reactant present.

Example

10 cm³ of 0.2 mol l⁻¹ sodium hydroxide solution is added to 25 cm³ of 0.1 mol l⁻¹ hydrochloric acid.

$NaOH + HCl \rightarrow NaCl + H_2O$

a Calculate which reactant is in excess.
b Calculate the mass of sodium chloride produced.

a • Step 1: Work out the key mole relationship from the balanced equation.

$NaOH + HCl \rightarrow NaCl + H_2O$

Key relationship: 1 mol   1 mol

- Step 2: Calculate the **number of moles** of each reactant from the information in the question.

  moles = concentration × volume

  NaOH: moles = 0·2 × 0·01

  = 0·0020 mol

  HCl: moles = 0·1 × 0·025

  = 0·0025 mol

  The sodium hydroxide and hydrochloric acid react in a 1:1 ratio, so the sodium hydroxide is the limiting reactant and the hydrochloric acid is in excess.

b The number of moles of sodium chloride can be calculated from the mole ratio in the balanced equation using the number of moles of the limiting reactant (sodium hydroxide).

- Step 1: Work out the key mole relationship from the balanced equation.

  $$NaOH + HCl \rightarrow NaCl + H_2O$$

  Key relationship: 1 mol   1 mol

  0·002   0·002

- Step 2: Calculate the mass of sodium chloride from the moles produced.

  mass = mol × GFM

  = 0·002 × 58·5

  = 0·117 g

1. 30 cm³ of 0·1 mol l⁻¹ of sulfuric acid is added to 50 cm³ of 0·15 mol l⁻¹ potassium hydroxide.

   $$2KOH + H_2SO_4 \rightarrow K_2SO_4 + 2H_2O$$

   a Calculate which reactant is in excess.
   b Calculate the mass of potassium sulfate produced.

2. 3·38 g of magnesium was added to 75 cm³ of 2 mol l⁻¹ of nitric acid.

   $$Mg + 2HNO_3 \rightarrow Mg(NO_3)_2 + H_2$$

   a Calculate the reactant in excess.
   b Calculate the mass of magnesium nitrate produced.

3. Hydrogen gas can be produced in the laboratory by reacting zinc with hydrochloric acid. 4·9 g of zinc was added to 200 cm³ of 0·1 mol l⁻¹ of hydrochloric acid.

   $$Zn + 2HCl \rightarrow ZnCl_2 + H_2$$

   a Calculate which reactant is in excess.
   b Calculate the mass of hydrogen produced.

Exercise 12G Calculating volumes of gas

Hint Molar volume (V_m) is the volume of gas occupied by 1 mole of any gas. The molar volume varies with temperature and pressure. Converting mass to volume and volume to mass can be done by converting to moles first.

Gram formula mass (GFM) ↔ 1 mole ↔ molar volume (V_m)

1 Calculate the volume occupied by 0·25 mol of methane gas at room temperature and pressure. Take $V_m = 24$ l mol^{-1}.

2 Calculate the number of moles in 6·7 l of carbon dioxide if the molar volume is 23·2 l mol^{-1}.

3 Calculate the volume occupied by 4·3 g of oxygen when the molar volume is 23·5 l mol^{-1}.

> **Hint** Volumes of gas reacting and produced can be worked out from balanced equations because 1 mol = V_m.

Example

Calculate the volume of carbon dioxide produced when 150 cm³ of propane gas is burned in excess oxygen.

$$C_3H_8(g) + 5O_2(g) \rightarrow 3CO_2(g) + 4H_2O(\ell)$$

Work out the key mole relationship from the balanced equation then convert to volumes. This can be done because, for any gas, 1 mol = V_m.

$$C_3H_8(g) + 5O_2(g) \rightarrow 3CO_2(g) + 4H_2O(\ell)$$

1 mol → 3 mol

So, 1 vol → 3 vol

So, 150 cm³ → 450 cm³

4 Calculate the volume of carbon dioxide produced when 250 cm³ of methane gas is burned in excess oxygen.

$$CH_4(g) + 2O_2(g) \rightarrow CO_2(g) + 2H_2O(\ell)$$

5 Calculate the volume of oxygen needed to react completely with 250 cm³ of butane.

$$C_4H_{10}(g) + \frac{13}{2}O_2(g) \rightarrow 4CO_2(g) + 5H_2O(\ell)$$

6 40 cm³ of propane was burned in 250 cm³ of oxygen.

$$C_3H_8(g) + 5O_2(g) \rightarrow 3CO_2(g) + 4H_2O(\ell)$$

a Show by calculation which reactant was in excess and by how much.

b Calculate the volume of carbon dioxide produced.

7 Calculate the volume of hydrogen produced when 4·7 g of magnesium reacts completely with excess hydrochloric acid. Take the molar volume (V_m) to be 23·9 l mol^{-1}.

$$Mg + 2HCl \rightarrow MgCl_2 + H_2$$

8 Nitrogen gas is produced in a car air bag by the rapid decomposition of sodium azide (NaN_3).

$$2NaN_3(s) \rightarrow 2Na(s) + 3N_2(s)$$

Calculate the volume of gas produced when 45 g of sodium azide decomposes if the molar volume (V_m) is 24·0 l mol^{-1}.

9 0·27 g of oxygen was found to occupy a volume of 200 cm³. Calculate the molar volume of oxygen.

13 Controlling the rate

Exercise 13A Collision theory

1 a Give **two** reasons why controlling the rate of reaction is important in industrial reactions.

b One factor which can affect reaction rate is the concentration of reactants. Name **four** other factors which can have an effect on the rate of a reaction.

2 Calculate the average rate of reaction, including units, given the following experimental results.

> **Hint** The relationship between the (average) reaction rate and the quantity of product produced over time is given in the SQA data booklet.

a $15\,cm^3$ of gas was collected between 10 and 30 seconds.

b The volume of gas collected increased from $10\,cm^3$ to $40\,cm^3$ between 0 and 50 seconds.

c The mass of magnesium decreased from $24.3\,g$ to $22.8\,g$ between 15 and 50 seconds.

d The volume of gas collected increased from $0.5\,l$ to $0.9\,l$ between 0 and 3 minutes.

3 The graph shows the volume of gas produced during the reaction of $40\,cm^3$ of $0.2\,mol\,l^{-1}$ hydrochloric acid reacting with excess large marble chips (calcium carbonate).

calcium carbonate + hydrochloric acid → calcium chloride + water + carbon dioxide

$CaCO_3(s) + 2HCl(aq) \rightarrow CaCl_2(aq) + H_2O(\ell) + CO_2(g)$

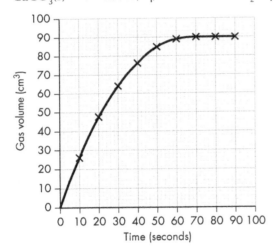

a Calculate the average rate of reaction over the first 10 seconds of the reaction.

b Sketch the graph (no need to include the scales) and label it Experiment 1.

Add curves to the graph to show results that would be obtained if the experiment was repeated using the conditions detailed below in part **i** (Experiment 2) and part **ii** (Experiment 3).

Label the curves obtained for Experiments 2 and 3.

(Pay particular attention to the slope of the graph and the final volume of gas produced.)

i Experiment 2: 40 cm³ of 0·2 mol l⁻¹ hydrochloric acid reacting with excess marble powder.

   ii Experiment 3: 20 cm³ of 0·2 mol l⁻¹ hydrochloric acid reacting with excess large marble chips.

> **Area 4 – Researching Chemistry:**
> Common chemical apparatus;
> General practical techniques
>
> c Draw a labelled diagram of the assembled apparatus which could be used to carry out this experiment.

> **Hint** The relationship between relative (reaction) rate and time is given in the SQA data booklet.

4 Strips of magnesium ribbon, with exactly the same mass, were dropped into beakers containing excess hydrochloric acid, but each with a different concentration. The time taken for the magnesium to react completely with each concentration of acid is shown in the table.

Concentration of hydrochloric acid (mol l⁻¹)	Time taken for all the magnesium to react (s)	Relative rate x 10³ (y)
0·10	470	2·1
0·15	345	2·9
0·20	250	4·0
0·25	200	5·0
0·30	172	

a State **two** observations which would indicate the reaction in each beaker had stopped.

b i Give the units for relative rate (y).

   ii Calculate the relative rate of reaction when the concentration is 0·30 mol l⁻¹. Include the units in your answer.

> **Area 4 – Researching Chemistry:**
> Reporting experimental work
>
> iii Draw a line graph of concentration against relative rate.

c i From the results state a conclusion that can be drawn about the relationship between concentration of acid and the rate of reaction.

   ii Explain, with reference to collision theory, how increasing the concentration of the acid affects the reaction rate.

5 The effect the concentration of reactants has on reaction rate can be shown using the reaction between hydrogen peroxide and an acidified solution of iodide ions.

$$H_2O_2(aq) + 2H^+(aq) + 2I^-(aq) \rightarrow 2H_2O(\ell) + I_2(aq)$$

If a small quantity of thiosulfate ions and starch are added, the thiosulfate ions will react with the iodine molecules as they form and they are changed back to iodide ions.

$$I_2(aq) + 2S_2O_3^{2-}(aq) \rightarrow 2I^-(aq) + S_4O_6^{2-}(aq)$$

Once all the thiosulfate has reacted, the iodine will now react with the starch and a colour change can be clearly seen. The time taken to reach this point in each reaction is noted and used to work out the relative rate of each reaction at different concentrations of iodide solution. The concentration of the iodide solution is varied by diluting the original concentration with water and keeping the total volume of iodide solution the same in each experiment.

The results are shown in the table.

Experiment	1	2	3	4	5
Volume of potassium iodide (cm³)	25	20	15	10	5
Volume of water (cm³)	0	5	(x)	15	20
Time (t) (s)	20	26	31	(y)	104
Rate (1/t) (s⁻¹)	0·050	0·038	0·032	0·019	0·010

a State the colour change which would be seen at the point the clock should be stopped.

b i State the volume of water (x) which should be added to the iodide solution in Experiment 3.

ii Explain your answer to part **i**.

iii Calculate a value for (y).

c Changing the concentration in this way allows the volume of iodide to be used as a measure of the concentration of the iodide ions.

A graph of relative rate against volume of iodide is shown.

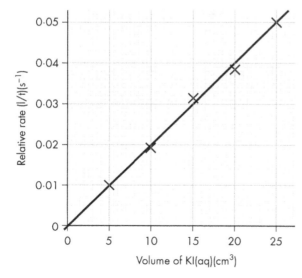

> **Area 4** – Researching Chemistry:
> Reporting experimental work
>
> **i** A line of best fit has been drawn through the points.
>
> Explain what is meant by a *line of best fit*.

    ii State the relationship between the relative rate of the reaction and the concentration of iodide ions.

    iii Calculate how long it would take for the colour change to take place when the relative rate is $0.025\,s^{-1}$.

d A student suggested adding different volumes of water to the original $25\,cm^3$ of iodide solution in order to change its concentration in each experiment.

Give **one** reason for not changing the concentration in this way.

6 The diagrams show how the rate of a reaction between hydrochloric acid and excess calcium carbonate can be monitored.

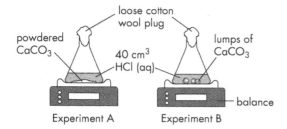

The graph shows how the total mass of the flask and contents changes with time in Experiment A.

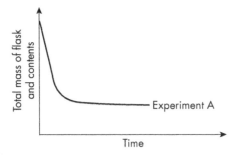

a Sketch the graph and add the curve which would be obtained from the results of Experiment B.

b Explain, with reference to collision theory, how decreasing the particle size of the calcium carbonate affects the reaction rate.

7 The molecular diagrams show how two reactant gas molecules might approach each other.

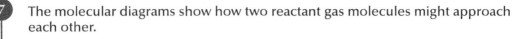

13 Controlling the rate

a i Suggest which arrangement, (a) or (b), is more likely to result in the products shown in the molecular diagram.

   ii Explain your answer to part **i**.

b Explain, with reference to collision theory, how increasing the pressure affects the reaction rate of gases.

Exercise 13B Reaction pathways

1 The diagram shows how the potential energy changes during a reaction.

a Copy the diagram and add the following:
   i   the enthalpy change, ΔH
   ii the activation energy, E_a
   iii where the activated complex, Y, forms.

b i   State what type of reaction is taking place.
   ii Explain your answer to part **i**.

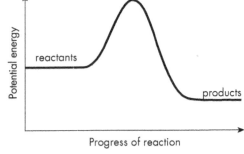

2 The diagram shows how the potential energy changes during a reaction.

a Copy the diagram and add the following:
   i   the enthalpy change, ΔH
   ii the activation energy, E_a
   iii where the activated complex, Y, forms.

b i   State what type of reaction is taking place.
   ii Explain your answer to part **i**.

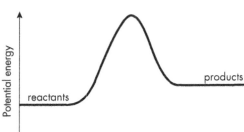

3 Iron is extracted from its ore in a blast furnace. Three main reactions take place.

Reaction 1: $C(s) + O_2(g) \rightarrow CO_2(g)$ $\Delta H = -394 \text{ kJ mol}^{-1}$
Reaction 2: $C(s) + CO_2(g) \rightarrow 2CO(g)$ $\Delta H = +173 \text{ kJ mol}^{-1}$
Reaction 3: $Fe_2O_3(s) + 3CO(g) \rightarrow 2Fe(\ell) + 3CO_2(g)$ $\Delta H = -23 \text{ kJ mol}^{-1}$

a i   State whether each reaction is exothermic or endothermic.
   ii Explain each of your answers in part **i**.

b The potential energy diagram for Reaction 1 is shown.
   i   Draw a similar potential energy diagram for Reaction 2.
   ii Draw a similar potential energy diagram for Reaction 3.

c In each of the potential energy diagrams you drew in part **b**, add:
   i   the activation energy
   ii the position of the activated complex, Y.

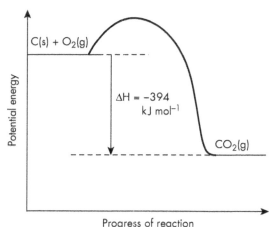

4 a Draw potential energy diagrams for the following reactions. (The diagrams do not need to be to scale.)

  i $H_2(g) + \frac{1}{2}O_2(g) \rightarrow H_2O(\ell)$ $\Delta H = -286\,kJ\,mol^{-1}$

  ii $C(s) + H_2O(g) \rightarrow CO(g) + H_2(g)$ $\Delta H = +121\,kJ\,mol^{-1}$

  b To each of the diagrams in part **a**, add:

  i the enthalpy change, ΔH

  ii the activation energy, E_a.

  c i State whether each reaction is exothermic or endothermic.

  ii With reference to the potential energy diagrams, explain your answers to part **i**.

5 The potential energy diagrams for three different reactions, A, B and C, are shown.

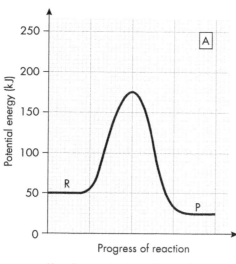

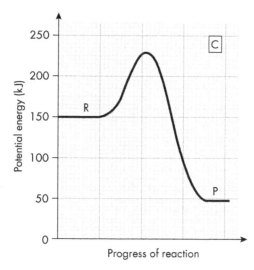

  a i State whether each potential energy diagram represents an exothermic or endothermic reaction.

  ii Explain each of your answers in part **i**.

  b From the information in each potential energy diagram calculate:

  i the enthalpy change for each reaction

  ii the activation energy for each reaction.

c i State which of the reactions is likely to be the fastest.

ii Explain your answer to part i.

6 The potential energy diagram is shown for a reversible reaction.

a Calculate the enthalpy change for:
  i the forward reaction
  ii the backward reaction.

b Calculate the activation energy for:
  i the forward reaction
  ii the backward reaction.

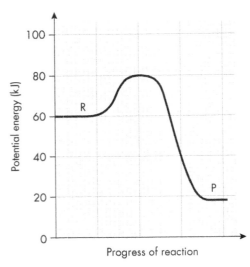

Exercise 13C The effect of catalysts

1 The potential energy diagram is for a reversible reaction.

a Draw out the diagram and add a curve to show what would happen when a catalyst is added.

b i Mark on your diagram the energy of activation for the catalysed and uncatalysed forward reactions.

ii Explain why adding a catalyst results in more reactants forming products.

c State the effect adding a catalyst has on the enthalpy change for:
  i the forward reaction
  ii the backward reaction.

2 Hydrogen peroxide (H_2O_2) slowly decomposes into water and oxygen. The enthalpy change (ΔH) for the reaction is $-25\,kJ\,mol^{-1}$ and the activation energy (E_a) is $75\,kJ\,mol^{-1}$.

a i State whether the reaction is exothermic or endothermic.

ii Explain your answer to part i.

b Copy and complete the potential energy diagram for the reaction (use graph paper).

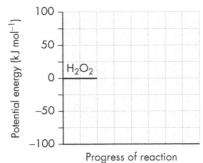

c Manganese dioxide powder catalyses the decomposition of hydrogen peroxide.
  i Add a curve to your potential energy diagram to show the path of the reaction when the catalyst is used.
  ii Explain why the hydrogen peroxide molecules decompose quicker when the catalyst is added.
  iii State why the catalyst is more effective when it is powdered rather than in lumps.

3 The potential energy diagram for a reversible reaction is shown. One curve is for the catalysed reaction and the other is for the uncatalysed reaction.

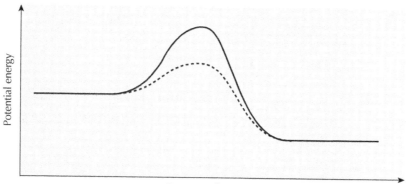

a Identify the curve for the catalysed reaction.
b State what effect the catalyst has on the activation energy (E_a) for:
  i the forward reaction
  ii the backward reaction.
c State what effect the catalyst has on the enthalpy change (ΔH) for:
  i the forward reaction
  ii the backward reaction.

4 The potential energy diagram below is for a reaction carried out with and without a catalyst.

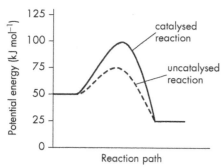

a Calculate the enthalpy change for the catalysed reaction.
b Calculate the activation energy for the uncatalysed reaction.

13 Controlling the rate

5 A catalytic converter in a car exhaust is made up of a honeycomb structure coated with a mixture of transition metals which acts as the catalyst. Pollutant gases are converted to less harmful gases. The diagram shows what happens.

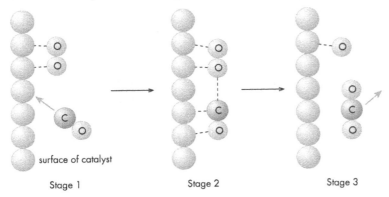

a Describe what is happening at each of the three stages.

b The honeycomb structure provides a large surface area for the catalyst.
State the advantage of this.

6 The reaction between potassium sodium tartrate (Rochelle's salt) and hydrogen peroxide (both colourless) can be catalysed using a solution of pink cobalt(II) ions. The colour of the reaction mixture immediately turns green and the reaction is at its fastest. As the reaction slows the pink colour reappears.

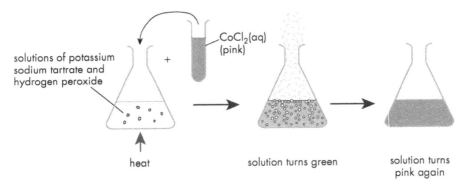

Use the information to answer the following.

a What evidence is there to indicate that the catalyst has taken part in the reaction?

b What evidence is there to indicate that the catalyst is regenerated at the end of the reaction?

c Suggest how you could tell, by observing the reaction, when it was at its fastest.

7 Hydrogenation is one of the most widely used industrial processes. One common example is the hardening of vegetable oils to make margarine. The oils and hydrogen gas are passed over a heated catalyst. Nickel is commonly used but involves high temperature and pressure, which is costly. Rare metals like rhodium and platinum are efficient catalysts but there are economic and environmental factors associated with their use. Iron can be used as a catalyst but only in the absence of water. To try and overcome this problem, chemists have developed iron nanoparticles inside a protective polymer structure. The technique is claimed to be safe, cheap, sustainable and environmentally friendly.

a Explain how catalysts enable products to form more quickly than when the catalyst isn't present.

b Suggest why a catalyst is more efficient in the form of nanoparticles than as a powder.

c i Suggest what one of the economic factors against using rare metals might be.

  ii Suggest what one of the environmental factors against using rare metals might be.

d i State what is likely to happen to iron in the presence of water and how this might affect it acting as a catalyst.

  ii State one property the polymer must have in order to prevent water reaching the iron.

  iii State why it is important for reactants to come into contact with the catalyst's surface.

e Suggest why using iron nanoparticles in a polymer is considered to be sustainable.

Exercise 13D Kinetic energy distribution

1 A group of students studied the effect of temperature on the rate of reaction using the reaction between oxalic acid and acidified potassium permanganate solution. They noted the time it took for the reaction to reach the same point at different temperatures. This was indicated by a sharp colour change.

a Suggest what colour change would take place.

b The students' results are shown in the graph.

  Use the graph to calculate the reaction time at:

  i 60°C

  ii 70°C.

c Explain why a small rise in temperature results in a large increase in the rate of a reaction.

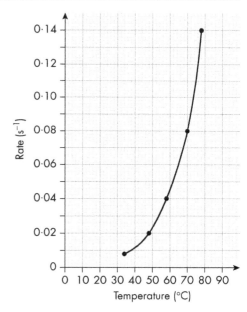

2 Explain the following observations.

a i When a Bunsen burner is switched on, but not lit, there is no observable evidence of a reaction taking place even though there are many collisions taking place between reactant molecules.

  ii When a flame or spark is brought close to the gas mixture in part i evidence that a reaction occurs quickly is clearly seen.

b As part of the Ostwald process, colourless nitrogen monoxide reacts with oxygen to form brown nitrogen dioxide. No input of energy is required for the reaction.

3 The diagram shows the energy distribution of molecules in a gas at a particular temperature.

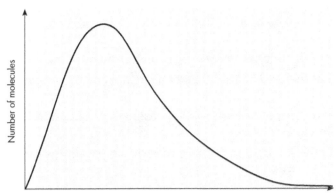

a State the relationship between temperature and the kinetic energy of the molecules.
b i Copy the graph and add a second curve to show the energy distribution of the molecules at a higher temperature.
ii Add a line to the graph to show where the activation energy (E_a) for the reaction would be.
iii Explain the position of E_a on your graph.

4 The energy distribution curve is for a gas at 50°C.

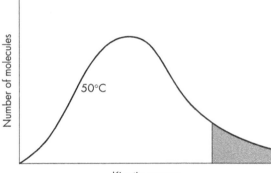

a State what the shaded area under the graph represents.
b Copy the graph and add a second curve to show the energy distribution of the molecules at 40°C.
c Mark on the graph the position of the activation energy (E_a).
d i State at which temperature the gas will be likely to react faster.
ii Explain your answer to part i.

5 a i Give the meaning of the term *enthalpy*.
ii Write the symbol for enthalpy change.
b The molecular diagrams represent reactants and products in a reaction. An unstable arrangement of atoms, **X**, is formed between reactants and products.

i Draw a possible shape for **X**.
ii State the term used to describe this unstable arrangement of atoms.
iii Explain the connection between the unstable arrangement of atoms and activation energy.

14 Chemical energy

Exercise 14A Measuring heat change by experiment

1 **a** **i** State the meaning of the term *enthalpy*.
   ii Write the symbol for enthalpy.
   iii Enthalpy cannot be measured but change in enthalpy can be. Give the symbol for change in enthalpy.

   b **i** Name the type of reaction in which energy is given out to the surroundings.
   ii Explain the importance of controlling industrial reactions in which heat is given out.
   iii Name the type of reaction in which energy is taken in from the surroundings.
   iv Give **one** reason that endothermic reactions in industry increase costs.

> **Hint** Enthalpy changes can be calculated indirectly by calculating the amount of heat energy transferred to a known mass of water during a chemical reaction. The relationship used is $E_h = cm\Delta T$, where:
>
> E_h = heat energy
>
> c = specific heat capacity of water (4.18 kJ kg^{-1} °C^{-1}; found in the SQA data booklet)
>
> m = mass of water (kg)
>
> ΔT = change in temperature
>
> This relationship is found in the SQA data booklet.
>
> By rearranging the relationship, c, m or ΔT can be calculated.

Example

Calculate the enthalpy change, in kJ mol^{-1}, when 0.75 g of sodium hydroxide is dissolved in 100 cm^3 of water. The rise in temperature was 3.9°C.

$c = 4.18$ $\quad E_h = cm\Delta T$

$m = 0.1$ $\quad\quad\quad = 4.18 \times 0.1 \times 3.9$

$\Delta T = 3.9$ $\quad E_h = 1.63$ kJ

Given the mass dissolved, the enthalpy change for 1 mol of sodium hydroxide dissolving can be calculated.

Moles dissolving = mass/GFM

$$= \frac{0.75}{40}$$

Moles dissolving = 0.019 mol

0.019 mol produces 1.63 kJ, so 1 mol would produce $\frac{1.63}{0.019} = 85.79$ kJ mol^{-1}.

Because there is a rise in temperature we know the reaction is exothermic, so the enthalpy change would normally be written as -85.79 kJ mol^{-1}.

2 When 3·4 g of potassium hydroxide is dissolved in 100 cm³ of water the rise in temperature is 6·9°C.

Calculate the enthalpy change, in kJ mol⁻¹, for potassium hydroxide dissolving.

3 The table shows the results of an experiment carried out to find the enthalpy change when potassium nitrate (KNO_3) is dissolved in water.

Measurement taken	Result
Mass of potassium nitrate (g)	2·9
Mass of water (g)	50·0
Initial temperature of water (°C)	20·3
Lowest temperature of water reached (°C)	17·8

Use this information to calculate the enthalpy change, in kJ mol⁻¹, when one mole of potassium nitrate is dissolved.

> **Hint** When calculating the enthalpy change when two solutions are reacted, the mass of water is taken as the total volume of the two solutions.

4 In an experiment, 50 cm³ of 1·0 mol l⁻¹ hydrochloric acid was added to 50 cm³ of 1·0 mol l⁻¹ of sodium hydroxide. The temperature rise was 6·1°C.

  a Calculate the heat energy, E_h, for the reaction, in kJ.

  b The equation for the reaction is:

  NaOH(aq) + HCl(aq) → NaCl(aq) + H_2O(ℓ)

  Calculate the enthalpy change, in kJ mol⁻¹, for 1 mole of water being produced.

5 A simple calorimeter used to determine the enthalpy of combustion is shown in the diagram.

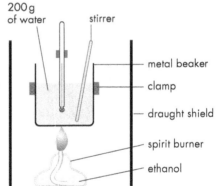

When 0·79 g of ethanol (C_2H_5OH) was burned the temperature of the water was raised by 21°C.

  a Calculate the enthalpy of combustion, ΔH_c in kJ mol⁻¹.

  b A data book value for the enthalpy of combustion of ethanol was given as −1367 kJ mol⁻¹.

  Explain why there is a significant difference between this value and the experimental value calculated in part **a**.

> **Area 4 – Researching Chemistry:** General practical techniques

  c Suggest why:

  i a metal can is used instead of a glass beaker

  ii an effort is made to reduce any draughts.

6 A group of students used a simple calorimeter to obtain data to calculate the enthalpy of combustion of methanol (CH_3OH). The results are shown in the table.

Measurement	Result
Mass of burner + methanol before burning (g)	52·17
Mass of burner + methanol after burning (g)	51·98
Mass of water heated (g)	100
Initial temperature of water (°C)	19
Highest temperature of water reached (°C)	29

Calculate the enthalpy of combustion of methanol.

7 It was found that the enthalpy change which occurred when 1·0 g of ammonium nitrate dissolved in 50 g of water was +0·33 kJ.

a Calculate the change in the temperature of the water.

b i State whether the reaction is exothermic or endothermic.

ii Explain your answer to part **i**.

8 It was found that when 0·8 g of sodium hydroxide dissolved in water the temperature rose by 4·8°C and the enthalpy change was −0·70 kJ.

Calculate the mass of water used in the experiment.

9 Data obtained from an experiment to find the specific heat capacity of a liquid is shown in the table.

Measurement	Result
Enthalpy change (kJ)	5·1
Mass of liquid heated (g)	100
Initial temperature of liquid (°C)	23·1
Final temperature of liquid (°C)	44·3

Calculate the specific heat capacity of the liquid, in $kJ\ kg^{-1}\ °C^{-1}$.

Exercise 14B Hess's law

> **Hint** Hess's law allows us to calculate enthalpies of reactions using enthalpy values obtained in data booklets.
>
> Enthalpies of formation and combustion can be found in the SQA data booklet.

1 Ethene can be hydrogenated to form ethane.

$C_2H_4(g) + H_2(g) \rightarrow C_2H_6(g)$

Calculate the enthalpy change for the reaction using the following enthalpies of combustion.

$C_2H_4(g) + 3O_2(g) \rightarrow 2CO_2(g) + 2H_2O(\ell)$ $\Delta H = -1411\ kJ\ mol^{-1}$

$C_2H_6(g) + \frac{7}{2}O_2(g) \rightarrow 2CO_2(g) + 3H_2O(\ell)$ $\Delta H = -1561\ kJ\ mol^{-1}$

$H_2(g) + \frac{1}{2}O_2(g) \rightarrow H_2O(\ell)$ $\Delta H = -286\ kJ\ mol^{-1}$

14 Chemical energy

2 The equation for the formation of ethanol is:

$$2C(s) + 3H_2(g) + \tfrac{1}{2}O_2(g) \rightarrow C_2H_5OH(\ell)$$

Use the following enthalpies of combustion to calculate the enthalpy of formation of methanol.

$C_2H_5OH(\ell) + 3\tfrac{1}{2}O_2(g) \rightarrow 2CO_2(g) + 3H_2O(\ell)$ $\Delta H = -1367 \text{ kJ mol}^{-1}$

$C(s) + O_2(g) \rightarrow CO_2(g)$ $\Delta H = -394 \text{ kJ mol}^{-1}$

$H_2(g) + \tfrac{1}{2}O_2(g) \rightarrow H_2O(\ell)$ $\Delta H = -286 \text{ kJ mol}^{-1}$

3 The equation for the combustion of ammonia is:

$$4NH_3(g) + 5O_2(g) \rightarrow 4NO(g) + 6H_2O(g)$$

Calculate the enthalpy change for the reaction using the following information.

$N_2(g) + O_2(g) \rightarrow 2NO(g)$ $\Delta H = -181 \text{ kJ mol}^{-1}$

$N_2(g) + 3H_2(g) \rightarrow 2NH_3(g)$ $\Delta H = -92 \text{ kJ mol}^{-1}$

$H_2(g) + \tfrac{1}{2}O_2(g) \rightarrow H_2O(\ell)$ $\Delta H = -286 \text{ kJ mol}^{-1}$

4 Use the enthalpy of formation values below to calculate the enthalpy of reaction shown by the following equation:

$$CH_4(g) + 2O_2(g) \rightarrow CO_2(g) + 2H_2O(\ell)$$

$C(s) + O_2(g) \rightarrow CO_2(g)$ $\Delta H_f(CO_2) = -394 \text{ kJ mol}^{-1}$

$H_2(g) + \tfrac{1}{2}O_2(g) \rightarrow H_2O(\ell)$ $\Delta H_f(H_2O) = -286 \text{ kJ mol}^{-1}$

$C(s) + 2H_2(g) \rightarrow CH_4(g)$ $\Delta H_f(CH_4) = -75 \text{ kJ mol}^{-1}$

5 Methylhydrazine (CH_3NHNH_2) can be used as a rocket fuel. The oxygen needed for combustion was provided by dinitrogen tetroxide (N_2O_4).

$$4CH_3NHNH_2(\ell) + 5N_2O_4(\ell) \rightarrow 4CO_2(g) + 12H_2O(\ell) + 9N_2(g)$$

Calculate the enthalpy of reaction from the enthalpy of formation of each of the reactants and products shown below.

$C(s) + 3H_2(g) + N_2(g) \rightarrow CH_3NHNH_2(\ell)$ $\Delta H_f = +54 \text{ kJ mol}^{-1}$

$N_2(g) + 2O_2(g) \rightarrow N_2O_4(\ell)$ $\Delta H_f = -20 \text{ kJ mol}^{-1}$

$C(s) + O_2(g) \rightarrow CO_2(g)$ $\Delta H_f = -394 \text{ kJ mol}^{-1}$

$H_2(g) + \tfrac{1}{2}O_2(g) \rightarrow H_2O(\ell)$ $\Delta H_f = -286 \text{ kJ mol}^{-1}$

6 The equation for the formation of methanol is shown.

$$C(s) + 2H_2(g) + \tfrac{1}{2}O_2(g) \rightarrow CH_3OH(\ell)$$

Calculate the enthalpy of formation of methanol from the enthalpies of combustion given below.

$\Delta H_c(CH_3OH(\ell)) = -726 \text{ kJ mol}^{-1}$

$\Delta H_c(C(s)) = -394 \text{ kJ mol}^{-1}$

$\Delta H_c(H_2(g)) = -286 \text{ kJ mol}^{-1}$

Exercise 14C Bond and mean bond enthalpies

> **Hint** Use the molar bond enthalpies and mean bond enthalpies in the SQA data booklet to answer these questions.

1 Explain the difference between *bond enthalpies* and *mean bond enthalpies*.

2 Use bond enthalpies and mean bond enthalpies to calculate the estimated enthalpy change for the following reaction.

$H_2(g) + Cl_2(g) \rightarrow 2HCl(g)$

3 Use bond enthalpies and mean bond enthalpies to calculate the estimated enthalpy change when hydrogen burns in excess oxygen.

$H_2(g) + \tfrac{1}{2}O_2(g) \rightarrow H_2O(g)$

4 Use bond enthalpies and mean bond enthalpies to calculate the estimated enthalpy change for the following reaction.

$N_2(g) + 3H_2(g) \rightarrow 2NH_3(g)$

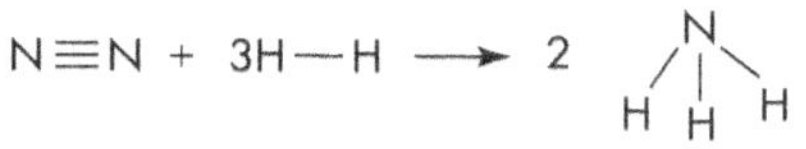

N≡N + 3H—H ⟶ 2 (NH₃ structure)

5 Use bond enthalpies and mean bond enthalpies to calculate the estimated enthalpy change when methane burns in excess oxygen.

$CH_4(g) + 2O_2(g) \rightarrow CO_2(g) + 2H_2O(g)$

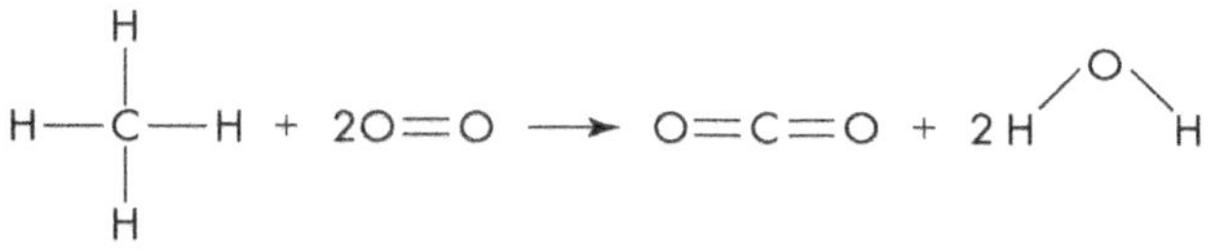

H—C(H)(H)—H + 2O=O ⟶ O=C=O + 2 H—O—H

6 Use bond enthalpies and mean bond enthalpies to calculate the estimated enthalpy change for the following reaction.

methanol + steam → hydrogen + carbon dioxide

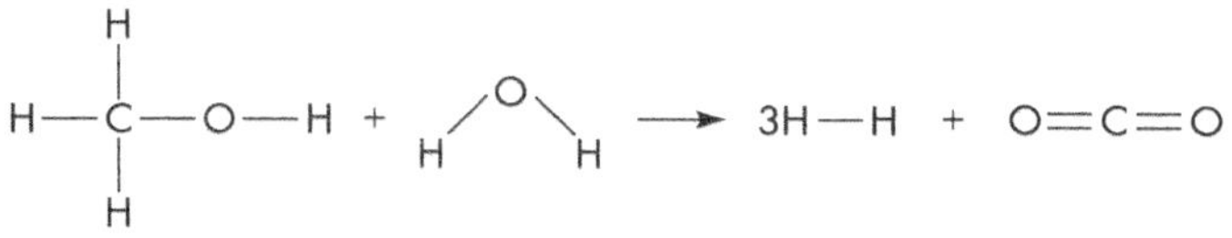

H—C(H)(H)—O—H + H—O—H ⟶ 3H—H + O=C=O

7 Methanamide (HCONH₂) is widely used in industry to make nitrogen compounds. It can be produced in the lab by reacting methanoic acid with ammonia.

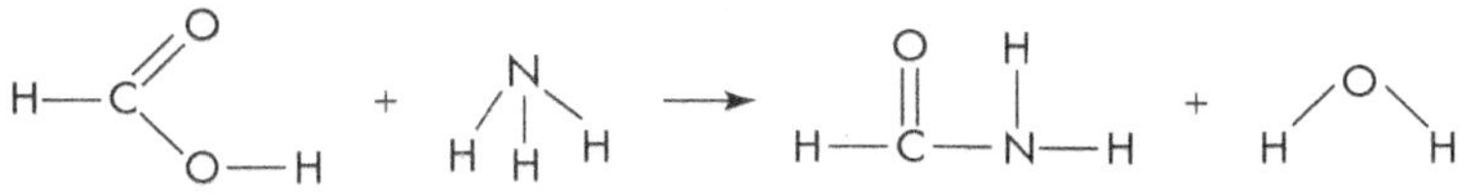

H—C(=O)—O—H + NH₃ ⟶ H—C(=O)—N(H)—H + H—O—H

Use bond enthalpies and mean bond enthalpies to calculate the estimated enthalpy change for the reaction. (The mean bond enthalpy of the C–N bond is 305 kJ mol⁻¹.)

15 Equilibria

Exercise 15A Reversible reactions and equilibrium

 Many reactions are reversible so products may be in equilibrium with reactants. This creates a number of issues for manufacturers of industrial products.

a i Explain what is meant by a *reversible reaction*.
   ii State what is meant by reactants and products being in *equilibrium*.
   iii The equilibrium is often described as dynamic.
       Explain what is meant by *dynamic* when describing equilibrium.
   iv Explain why a state of dynamic equilibrium can only be attained in a closed system.

b Give **two** reasons why the formation of an equilibrium in a chemical reaction is a problem for industrial chemists.

 A reversible reaction at equilibrium can be represented as follows.

A + B ⇌ C + D

a The graph shows how the rates of the forward and backward reactions change over time.

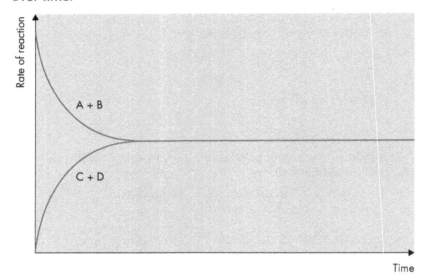

   i Explain the rate of the forward reaction at the start of the reaction.
   ii Explain what happens to the rate of the forward reaction as the reaction progresses.
   iii Explain the rate of the backward reaction at the start of the reaction.
   iv Explain what happens to the rate of the backward reaction as the reaction progresses.

b At which point is a state of dynamic equilibrium reached?

c The graph below shows how the concentrations of the reactants and products change over time.

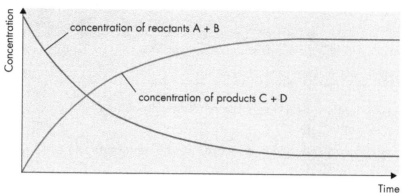

  i Explain what happens to the concentration of reactants A and B as the reaction progresses.
  ii Explain what happens to the concentration of C and D as the reaction progresses.
d At which point on the graph is equilibrium reached?
e Write a statement linking concentrations of reactants and products at equilibrium for this reaction.
f i State whether this reaction lies to the left or the right.
  ii Explain your answer to part **i**.

3 The German chemist Max Ernst Bodenstein studied the reaction between hydrogen and iodine.

$$H_2(g) + I_2(g) \rightleftharpoons 2HI(g)$$

Hydrogen and iodine mixtures (**B**) were sealed in glass containers and kept at the same temperature. The same procedure was carried out with a container of hydrogen iodide (**A**). The flasks were cooled at certain time intervals to stop the reactions in the containers and the number of moles of hydrogen iodide in flasks **A** and **B** was analysed. The results of the experiments are shown in the graph.

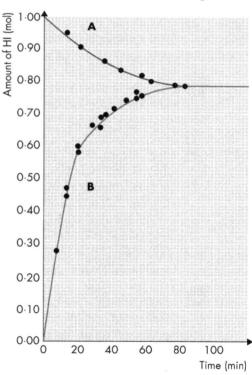

a i Give the number of moles of hydrogen iodide in flask **A** at the start of the experiment.
  ii Give the number of moles of hydrogen iodide in flask **B** at the start of the experiment.
  iii Give the number of moles of hydrogen iodide in flasks **A** and **B** at equilibrium.
  iv Explain how you deduced your answer to part **iii**.

15 Equilibria

b Suggest why more than one flask of each of **A** and **B** was set up and tested.

> **Area 4** – Researching Chemistry: Reporting experimental work
>
> c A line of best fit is drawn through each set of results.
> State what is meant by *line of best fit*.

4 The general equation for a reversible reaction can be written as:

A + B ⇌ C + D

The graphs show the change in concentration of reactants and products over time.

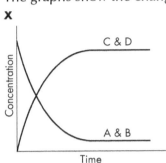
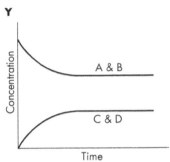

a i State whether the equilibrium position of the reaction represented by graph **X** lies to the right or left.

ii Explain your answer to part **i**.

b i State whether the equilibrium position of the reaction represented by graph **Y** lies to the right or left.

ii Explain your answer to part **i**.

c State at which point on the graphs equilibrium is reached.

5 Two test tubes, **X** and **Z**, are set up as shown in the diagram.

Test tube **X** shows potassium iodide solution on top of a solution of iodine in trichloroethane ($C_2H_3Cl_3$). Test tube **Z** shows iodine dissolved in potassium iodide solution on top of trichloroethane. When the two test tubes are shaken and the contents allowed to settle, both look like the contents of test tube **Y**.

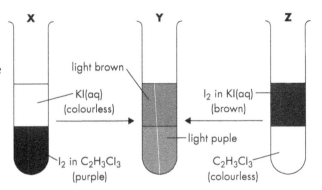

Explain fully this observation and any conclusions which can be drawn.

6 State whether each of the following statements is always true about a chemical reaction in a state of dynamic equilibrium.

a The concentrations of reactants and products are the same.

b The activation energies of the forward and backward reactions are equal.

c The rate of the forward and backward reactions are the same.

d No enthalpy change is involved during the reaction.

Exercise 15B Changing the position of equilibrium

1 The equilibrium formed when iron(III) ions (Fe^{3+}) and thiocyanate ions (CNS^-) react is summarised in the equation below.

Fe^{3+}(aq) + CNS^-(aq) ⇌ $[FeCNS]^{2+}$(aq)
pale yellow   colourless   red

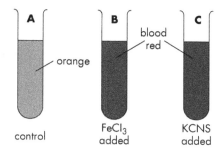

The two reactants were mixed and diluted with water until an orange colour was produced. The orange solution was divided equally between four test tubes, labelled **A–D**. One test tube, **A**, was kept as a control. Test tube **B** had a small volume of iron(III) chloride added. Test tube **C** had a small volume of potassium cyanide added. The colour changes which occurred are shown in the diagram.

  a Explain the colour change which took place in test tube **B**.

  b Explain the colour change which took place in test tube **C**.

  c Sodium chloride was added to test tube **D**. This acts to remove Fe^{3+}(aq) ions from solution.

   i Suggest what would happen to the colour in test tube **D**.

   ii Explain your answer to part **i**.

2 Esters are made by reacting a carboxylic acid with an alcohol. The reaction is reversible and an equilibrium is established.

  carboxylic acid + alcohol ⇌ ester + water

  a When an ester is made industrially a number of things are done to improve the yield of ester.

   i Water is removed as the ester is formed.

   Explain how this improves the yield of ester.

   ii Excess of one of the reactants is continually added.

   Explain how this improves the yield of ester.

  b A catalyst can be used in the reaction.

   State the effect adding a catalyst has on the equilibrium position in a reaction.

  c The graph shows how the concentrations of reactants and products change during this reaction.

   Write a statement about the concentrations of reactants and products at equilibrium.

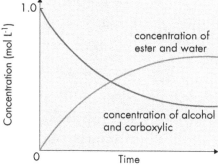

3 In the industrial preparation of methanol a mixture of carbon monoxide and hydrogen is passed over a catalyst.

  CO(g) + $2H_2$(g) ⇌ CH_3OH(g)

  a i Suggest what could be done to the pressure to increase the yield of methanol.

   ii Explain your answer to part **i**.

  b The yield of methanol is increased by liquefying it as it forms.

   Explain how this increases the yield.

  c State the effect the catalyst has on the equilibrium position.

4 a Predict the effect that increasing the pressure would have on the following equilibria.
   i $2CO(g) + O_2(g) \rightleftharpoons 2CO_2(g)$
   ii $H_2(g) + Cl_2(g) \rightleftharpoons 2HCl(g)$
   iii $2NO(g) + O_2(g) \rightleftharpoons 2NO_2(g)$
   iv $C_2H_6(g) \rightleftharpoons C_2H_4(g) + H_2(g)$
   b Explain each of your answers in part a.

5 Nitrogen dioxide (NO_2) is a brown gas which exists in equilibrium with dinitrogen tetroxide (N_2O_4), which is colourless.
   $NO_2(g) \rightleftharpoons 2N_2O_4(g)$ $\Delta H_{forward}$ = negative
   brown   colourless
   a i State what would happen to the colour of the gas mixture if the pressure was increased.
     ii Explain your answer to part i.
     iii State what would happen to the colour of the gas mixture if the pressure was then decreased.
     iv Explain your answer to part iii.
   b i State how an increase in temperature affects the equilibrium position.
     ii Explain your answer to part i.

6 Synthesis gas, a mixture of carbon monoxide and hydrogen, is used industrially in the production of methanol. It can be produced by passing steam and methane over a hot catalyst.
   $CH_4(g) + H_2O(g) \rightleftharpoons CO(g) + 3H_2(g)$ $\Delta H_{forward}$ = +206 kJ mol^{-1}
   a i Predict what effect raising the temperature would have on the yield of synthesis gas.
     ii Explain your answer to part i.
   b i Suggest what effect raising the pressure would have on the yield of synthesis gas.
     ii Explain your answer to part i.
   c Using a catalyst has no effect on the amount of synthesis gas produced.
   Suggest then why a catalyst is used.

7 State how the changes detailed in parts i–iii below would affect the concentration of NO.
   $N_2(g) + O_2(g) \rightleftharpoons 2NO(g)$ $\Delta H_{forward}$ = +180 kJ mol^{-1}
   a i Increasing the temperature.
     ii Increasing the pressure.
     iii Adding a catalyst.
   b Explain each of your answers to part a.

8 The following questions relate to the reaction:
   $4NH_3(g) + 5O_2(g) \rightleftharpoons 4NO(g) + 6H_2O(g)$ $\Delta H_{forward}$ = negative
   State the effect the following changes would have on the position of equilibrium.
   a Increasing the temperature.
   b Increasing the pressure.
   c Adding a catalyst.
   d Increasing the concentration of NH_3.
   e Removing NO.

16 Chemical analysis

Exercise 16A Chromatography

 Chromatography is a method of separating the components in a mixture. All forms of chromatography operate on the same principle. The mixture is introduced into two phases – a stationary phase and a mobile phase.

The diagram shows an example of the apparatus used in paper chromatography and a resulting chromatogram. A, B and C are mixtures of compounds. The stationary phase is the chromatography paper.

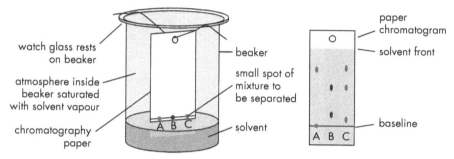

a State the **two** properties of substances which affect the rate at which they are separated from a mixture using chromatography.

b i Suggest what is meant by the mobile phase in the paper chromatography example shown.

ii Suggest why a watch glass is placed on top of the beaker and the atmosphere inside the beaker is saturated with solvent vapour.

c i The compounds in a mixture can be identified by their retention factors.
   Explain what is meant by *retention factor*.

ii Suggest why mixture A shows only one spot above the baseline.

d Apart from separating substances, suggest **one** other benefit of using paper chromatography in chemical analysis.

 Paper chromatography can be used to separate mixtures of amino acids. The retention factor (R_f) can be worked out from the resulting chromatogram.

$$R_f = \frac{\text{distance moved by the substance}}{\text{maximum distance moved by the solvent}}$$

The diagram below shows how a chromatogram can be obtained.

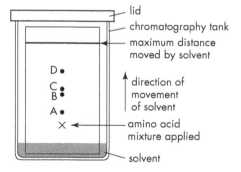

a i Calculate the R_f value for each of the substances in the mixture.

  ii State how retention factors can be used.

b Water on the surface of the paper acts as the solvent in the stationary phase. The solvent which moves up the paper is known as the mobile phase.

  i Suggest which of the amino acids in the chromatogram is most soluble in the stationary phase.

  ii Suggest why polar substances might spend more time in the stationary phase than in the mobile phase if hexane is used as the solvent in the mobile phase.

3 The gas–liquid chromatogram shows the additives in a soft drink.

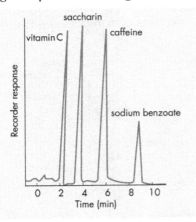

a i Estimate which of the additives there is most of.

  ii Explain your answer to part **i**.

b i State which of the additives is carried through the column fastest.

  ii State the term used for the time it takes for a component in a mixture to travel through the apparatus.

c A chromatogram obtained from another sample of a soft drink, which manufacturers claimed was the same drink but caffeine-free, is shown.

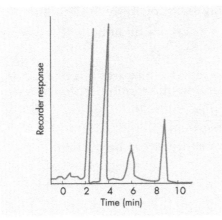

Comment on the validity of their claim.

Exercise 16B Volumetric analysis

1 Volumetric analysis is a technique which involves the titration of a **standard solution** with a solution of unknown concentration. The volume of standard solution needed for complete reaction is obtained by averaging **concordant titres**. An **indicator** is usually added to enable the **end-point** to be measured.

a Give the meaning of each of the terms in bold.

b Sodium carbonate can be used as a standard solution in an acid–base titration.

  i Calculate the mass of sodium carbonate (Na_2CO_3) required to make 100 cm³ of 0·10 mol l⁻¹ sodium carbonate solution.

> **Area 4** – Researching Chemistry: General practical techniques
>
> **ii** Outline how you would make the standard solution in part **i**, using the mass of sodium carbonate you calculated.

2 The concentration of an iron(II) sulfate solution was calculated by titrating 20·0 cm³ samples with acidified potassium permanganate solution (0·20 mol l⁻¹). The results are shown in the table.

Titre	Volume of acidified permanganate added (cm³)
1	24·8
2	23·9
3	24·1

a Explain why an indicator does not need to be added in this titration.

b State why the permanganate solution is acidified.

> **Area 4** – Researching Chemistry: Common chemical apparatus
>
> **c** Name the piece of apparatus used to measure exactly 20·0 cm³ of iron(II) sulfate solution.

> **Area 4** – Researching Chemistry: Reporting experimental work
>
> **d i** State the volume of permanganate solution which should be used to calculate the concentration of the iron(II) sulfate solution.

  ii Calculate the concentration of the iron(II) sulfate solution, using the volume of potassium permanganate from part **i**.

  Redox equation:

  $MnO_4^{2-}(aq) + 8H^+(aq) + 5Fe^{2+}(aq) \rightarrow Mn^{2+}(aq) + 5Fe^{3+}(aq) + 4H_2O(\ell)$

3 15·0 cm³ of a tin(II) chloride solution reacts with 20 cm³ of 0·2 mol l⁻¹ potassium dichromate solution which has been acidified with dilute sulfuric acid.

Redox equation:

$3Sn^{2+}(aq) + Cr_2O_7^{2-} + 14H^+(aq) \rightarrow 3Sn^{3+}(aq) + 2Cr^{3+}(aq) + 7H_2O(\ell)$

Calculate the concentration of the tin(II) chloride solution.

4 A vitamin C tablet ($C_6H_8O_6$; GFM = 176 g) was dissolved in some deionised water and made up to 250 cm³ in a volumetric flask. 20 cm³ of the solution was titrated with 0·025 mol l⁻¹ iodine solution. The average titre was 17·9 cm³.

Redox equation: $C_6H_8O_6 + I_2 \rightarrow C_6H_6O_6 + 2I^- + 2H^+$

a i Calculate the mass of vitamin C in the 20·0 cm³.

> **Hint** Calculate the number of moles of vitamin C in the sample and convert to mass.

ii Calculate the mass of vitamin C in the tablet.

b Suggest why deionised water is used to make up the vitamin C solution.

Higher
CHEMISTRY

**Mixed Exam
Question Practice**

Barry McBride

MULTIPLE-CHOICE QUESTIONS

1 Particles with the same electron arrangement are said to be isoelectronic.

Which of the following compounds contains ions that are isoelectronic?

A Li_2O

B NaCl

C K_2O

D $CaCl_2$

2 Which of the following elements is the least electronegative?

A Fluorine

B Chlorine

C Sodium

D Caesium

3 Which of the following reactions represents the first ionisation energy of chlorine?

A $Cl_2(\ell) + e^- \rightarrow Cl^-(g)$

B $Cl_2(g) + 2e^- \rightarrow 2Cl^-(g)$

C $Cl(g) \rightarrow Cl^+(g) + e^-$

D $Cl(g) \rightarrow Cl^{2+}(g) + 2e^-$

4 The difference in covalent radius between lithium and fluorine is due to the

A number of energy levels

B number of neutrons

C number of protons

D mass of each atom.

5 Which of the following is most likely to act as a reducing agent?

A $Cr_2O_7^{2-}$

B MnO_4^-

C H_2O_2

D CO

6 Which of the following is an isomer of 3,3-dimethylbutan-1-ol?

A $CH_3CH_2CH(CH_3)CH_2OH$

B $CH_3CH(CH_3)CH(CH_3)CH_2OH$

C $CH_3CH_2CH(CH_3)CH_2OH$

D $CH_3CH(CH_3)C(CH_3)_2CH_2OH$

7 Which of the following would not react with acidified potassium dichromate solution?

A Butan-2-ol

B Butanone

C Methanol

D Methanal

8 The compound shown is an example of

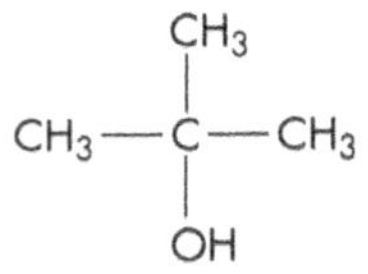

$$CH_3-\underset{\underset{OH}{|}}{\overset{\overset{CH_3}{|}}{C}}-CH_3$$

A a primary alcohol

B a secondary alcohol

C a tertiary alcohol

D an aldehyde.

9 Pineapple flavouring is based on the compound shown.

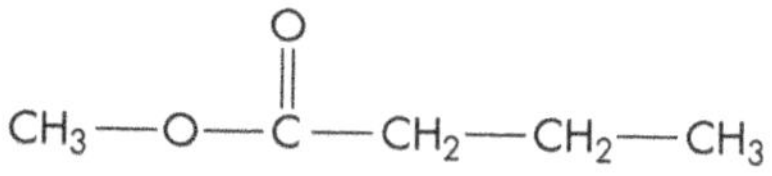

$$CH_3-O-\underset{\underset{O}{\|}}{C}-CH_2-CH_2-CH_3$$

It can be produced from

A methanol and butanoic acid

B propanol and ethanoic acid

C butanol and methanoic acid

D propanol and propanoic acid.

10 When an egg is heated, the protein it contains is denatured, causing it to change colour from colourless to white.

During denaturing, the protein molecule

A is oxidised

B is hydrolysed

C is reduced

D changes shape.

11 Glycerol contains

A no hydroxyl groups

B 1 hydroxyl group

C 2 hydroxyl groups

D 3 hydroxyl groups.

12 The reaction that takes place when vegetable oils are converted to vegetable fats is known as

A hydrolysis

B condensation

C hydrogenation

D dehydrogenation.

13 Emulsifiers are commonly added to food to

A prevent oxidation

B reduce the calorific value

C reduce the saturated fat content

D prevent oil and water molecules separating into layers.

14 The number of moles of ions in 1 mol of copper(II) sulfate is

A 1

B 2

C 3

D 4.

15 Which of the following gas samples has the same volume as 8.0 g of methane, CH_4?

(All volumes are measured at the same temperature and pressure.)

A 1.0 g of helium

B 2.0 g of hydrogen

C 8.0 g of oxygen

D 38.0 g of fluorine

16 Copper can be produced from copper(I) sulfide.

Cu_2S + O_2 → $2Cu$ + SO_2

GFM = 159.1 g   GFM = 32 g   GFM = 63.5 g   GFM = 64.1 g

The atom economy for the production of copper is

A 66.8%

B 66.5%

C 39.9%

D 33.2%

17 Which of the following correctly states the effect that adding a catalyst will have on a reaction mixture at equilibrium?

    A The position of equilibrium is unchanged.

    B The ΔH of the reverse reaction will increase.

    C The ΔH of the forward reaction will increase.

    D The position of equilibrium always shifts to the right.

18 The potential energy diagram for a reaction is shown.

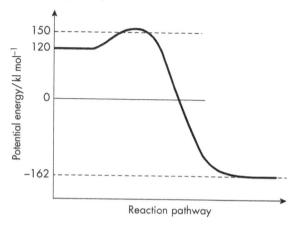

The enthalpy change for the forward reaction, in kJ mol⁻¹, is

    A −42

    B 155

    C −282

    D −182.

19

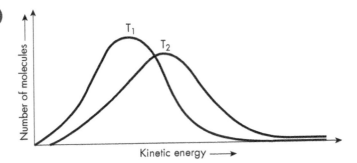

Which line in the table correctly interprets the above energy distribution diagram as the temperature increases from T_1 to T_2?

	Activation energy (E_a)	Number of successful collisions
A	Unchanged	Increases
B	Unchanged	Decreases
C	Increases	Increases
D	Increases	Decreases

20 The three equations shown below all involve displacement reactions between metals and metal oxides.

Mg(s) + FeO(s) → MgO(s) + Fe(s) ΔH = **A** kJ mol⁻¹

Fe(s) + CuO(s) → FeO(s) + Cu(s) ΔH = **B** kJ mol⁻¹

Mg(s) + CuO(s) → MgO(s) + Cu(s) ΔH = **C** kJ mol⁻¹

Which is the correct relationship between A, B and C according to Hess's Law?

A A + B = −C

B A + B = C

C C + A = −B

D C + A = B

21 One method used to produce methanol requires synthesis gas. The following equation shows the production of methanol from synthesis gas.

$$2H_2(g) + CO(g) \rightleftharpoons CH_3OH(g) \quad \Delta H = -91 \text{ kjmol}^{-1}$$

Which line in the table shows the conditions that would cause the greatest increase in the amount of methanol produced?

	Pressure	Temperature
A	High	High
B	Low	Low
C	High	Low
D	Low	High

22 Potassium chlorate can undergo thermal decomposition to produce potassium chloride and oxygen gas.

$$2KClO_3(s) \rightarrow 2KCl(s) + 3O_2(g)$$

What volume of oxygen would be obtained by the decomposition of 0.05 moles of potassium chlorate in such a reaction?

(The molar volume of oxygen under these conditions is 24 litres mol^{-1}.)

A 0.15 litres

B 0.3 litres

C 1.8 litres

D 3.6 litres

23 Xenon trioxide is an unstable compound that breaks down to form xenon gas.

$$XeO_3 + \_\_H^+ + \_\_e^- \rightarrow Xe + \_\_H_2O$$

The numbers of H$^+$, e$^-$ and H$_2$O required to balance this equation are

A 6H$^+$, 6e$^-$, 3H$_2$O

B 6H$^+$, 6e$^-$, 6H$_2$O

C 3H$^+$, 3e$^-$, 3H$_2$O

D 3H$^+$, 3e$^-$, 6H$_2$O.

24 A solution of accurately known concentration is more commonly known as a

  A correct solution

  B precise solution

  C prepared solution

  D standard solution.

25 The average kinetic energy of the particles in a substance can be measured using a

  A thermometer

  B balance

  C pipette

  D standard flask.

EXTENDED RESPONSE QUESTIONS

1 The structure of a fat molecule is shown

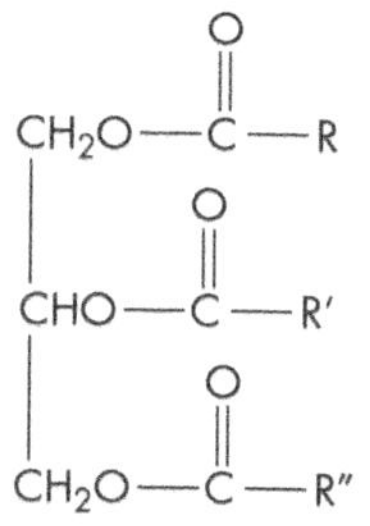

$$
\begin{array}{l}
CH_2O-\overset{O}{\underset{\|}{C}}-R \\
CHO-\overset{O}{\underset{\|}{C}}-R' \\
CH_2O-\overset{O}{\underset{\|}{C}}-R''
\end{array}
$$

a i When the fat is broken down, fatty acids are obtained.
The fatty acids are represented by R, R' and R" in the diagram.
State the name of the other product of this reaction.   1

ii State the name of the reaction that takes place when fats are broken down to form fatty acids.   1

b Fats are solid at room temperature, but oils are liquid.

Explain fully why fats have a higher melting point than oils.   2

c State the type of compound that fats and oils can be classified as.   1

d Paper chromatography can also be used to identify the products of this reaction.

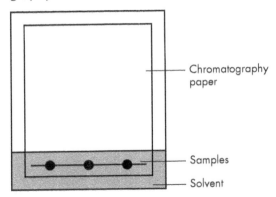

i State a factor that affects how far a sample travels up the chromatography paper. | 1

ii Suggest why setting up the chromatography experiment as shown in the diagram is incorrect. | 1

Total marks | 7

2 Copper metal can react with concentrated sulfuric acid to produce copper sulfate, water and the very soluble gas, sulfur dioxide.

$$Cu(s) + 2H_2SO_4(aq) \rightarrow SO_2(g) + CuSO_4(aq) + 2H_2O(\ell)$$

a i Complete the diagram to show how the sulfur dioxide produced in the experiment could be collected and the volume measured. | 1

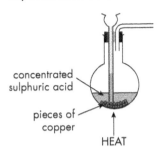

ii Sulfur dioxide has the structure shown

Explain fully why sulfur dioxide is very soluble in water. | 2

iii When the reaction was complete, 20 cm³ of sulfur dioxide was collected. Calculate the mass, in grams, of copper that reacted.
(Take molar volume to be 24 litres mol⁻¹)
Show your working clearly. | 2

b The potential energy diagram for the reaction is shown.

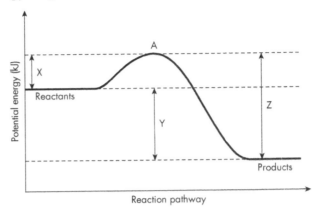

i State what could be used to lower the values of X and Z but will have no effect on Y. — 1

ii At point A, an unstable arrangement of atoms is formed.

State the name given to this arrangement. — 1

iii Potential energy diagrams can be used to calculate the activation energy for a reaction.

State what is meant by the term *activation energy*. — 1

Total marks 8

3 The combustion reactions of ethane and ethanol can be studied in different ways.

a The enthalpy of combustion of ethanol can be measured using a bomb calorimeter like the one shown.

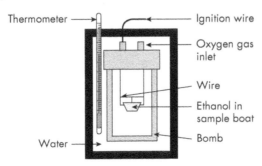

i When the experiment was performed in the calorimeter, it was found that 0.92 g of ethanol resulted in the temperature of 400 cm³ of water increasing from 18.2°C to 34.3°C.

Calculate the enthalpy of combustion of ethanol.

Show your working clearly.

3

ii The result obtained for the enthalpy of combustion of ethanol using the bomb calorimeter is higher than those obtained in the school lab.

Suggest a reason for this.

1

b The combustion of ethane produces carbon dioxide and water vapour.

$$C_2H_4(g) + 3O_2(g) \rightarrow 2CO_2(g) + 2H_2O(g)$$

i Using bond enthalpies and mean bond enthalpies from the data booklet, calculate the enthalpy change, in kJ mol⁻¹, for this reaction.

2

ii Explain the difference between bond enthalpy and mean bond enthalpy.

1

Total marks 7

4 Green salt, uranium tetrafluoride, is used to produce fuel for nuclear power stations. It is produced from uranium ore.

a Uranium can be extracted from green salt in a redox reaction with magnesium metal.

$$2Mg + UF_4 \rightarrow 2MgF_2 + U$$

i Name the oxidising agent in this reaction.

ii Calculate the atom economy for the reaction.

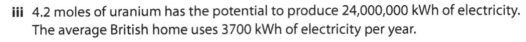

iii 4.2 moles of uranium has the potential to produce 24,000,000 kWh of electricity. The average British home uses 3700 kWh of electricity per year.

Calculate the mass of uranium required to power an average British home for a year.

b For this reaction to take place, the uranium tetrafluoride must be in the molten state; therefore, the reaction is carried out at a temperature of over 1100°C in an argon atmosphere.

Suggest why the reaction is not carried out in air. **1**

c Uranium hexafluoride is a solid with a melting point of 64.8°C.

Suggest the structure and bonding found in uranium hexafluoride. **2**

Total marks 8

5 August Kekulé was an organic chemist born in Germany in 1829. His work involved predicting the structure of organic compounds. He predicted the structure of benzene, C_6H_6, which is shown.

a It was later proven that benzene is not unsaturated.

Describe a chemical test, including the result, to show that benzene is not unsaturated. **2**

Extended response questions   117

b The equation for the formation of benzene from carbon and hydrogen is shown.

$$6C(s) + 3H_2(g) \rightarrow C_6H_6(\ell)$$

Use the enthalpies of combustion from your data book to calculate the enthalpy, in kJ mol^{-1}, of the formation of benzene

Show your working clearly.

3

c The solubility of benzene was also tested using several solvents including water and carbon tetrachloride.

Suggest in which of these solvents benzene would be more soluble, and explain your answer.

2

d Early Grand Prix cars were fuelled by a mixture of chemicals and additives that included large quantities of benzene, alcohols and aviation fuel.

The structure of one of the hydrocarbons found in aviation fuel is shown.

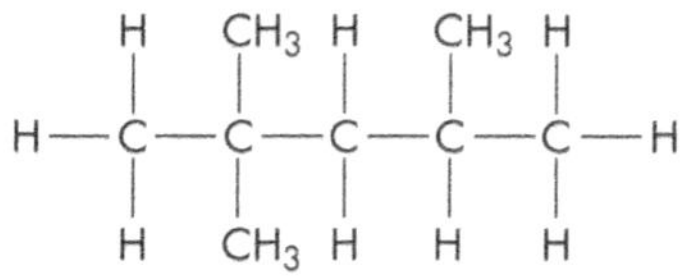

i Give the systematic name of this hydrocarbon.

1

ii Some modern cars are powered by fuel cells. Fuel cells generate electricity by chemical means.

The ion-electron equations, which occur at each electrode, are shown.

$$H_2(g) \rightarrow 2H^+(aq) + 2e^-$$

$$O_2(g) + 4H^+ + 4e^- \rightarrow 2H_2O(\ell)$$

Combine the two ion-electron equations to give the overall redox equation. **1**

iii Suggest why fuel cells are thought to be more environmentally friendly than aviation fuels. **1**

Total marks 10

6 The electrostatic force of attraction between opposite electrical charges is fundamental to chemistry.

Using your knowledge of chemistry, comment on why this force is so important in chemistry. **3**

7 Diphosphane is a non-polar compound with the formula P_2H_4.

a The compound is non-polar because the elements present in the compound have the same electronegativity.

    ii State what is meant by the term electronegativity.

    ii Draw a possible structure of diphosphane.

b Diphosphane is spontaneously flammable in air.

The equation for the combustion of diphosphane is

$$2P_2H_4(g) + 7O_2(g) \rightarrow P_4O_{10}(s) + 4H_2O(\ell)$$

    i Calculate the volume of oxygen, in cm³, that would be required to completely burn 30 cm³ of diphosphane.

    Show your working clearly.

    ii Suggest why it can be concluded that the activation energy for this reaction is low.

Total marks 4

Extended response questions

8 Iodine is slightly soluble in water. When excess iodine is mixed with water the following equilibrium is established.

$$I_2(s) + aq \rightleftharpoons I_2(aq) \quad \Delta H \text{ positive}$$

The concentration of dissolved iodine was measured over a period of time and the results plotted on a graph.

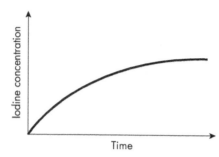

a Add a curve to show how the iodine concentration would change with time if the measurements were repeated at a higher temperature. **1**

b The dissolved iodine reacts with water as follows.

$$I_2(aq) + H_2O(\ell) \rightleftharpoons 2H^+(aq) + I^-(aq) + IO^-(aq)$$

Complete the table to show the effect of adding each of the solids on the position of equilibrium.

Solid added	Effect on equilibrium position
Potassium iodide	
Sodium hydroxide	

2

c i Iodine solution reacts with vitamin C according to the following equation

$$C_6H_8O_6(aq) + I_2(aq) \longrightarrow C_6H_6O_6(aq) + 2H^+(aq) + 2I^-(aq)$$
(brown) (colourless)

A student performed a titration with 50 cm³ of 0.1 mol l⁻¹ of vitamin C solution being added to 0.54 g of iodine in solution.

By calculating which reactant is in excess, state whether the iodine solution would have been decolourised.

Show your working clearly. 3

ii Apart from taking accurate measurements, suggest two points of good practice that a student should follow to ensure that an accurate end-point is achieved in a titration. 2

Total marks 8

9 Methane reacts with chlorine in sunlight to produce a mixture of chloroalkanes in an example of a free radical chain reaction.

a The sunlight is required to split the chlorine molecules to form free radicals.

  i Name the type of radiation, present in sunlight, that provides enough energy to split the chlorine molecules. — 1

  ii Write the equation for this reaction. — 1

b State what is meant by the term *free radical*. — 1

c State the name of the first stage in a free radical chain reaction. — 1

d Many cosmetic products contain free radical scavengers.

  i State what is meant by the term *free radical scavenger*. — 1

  ii Name another type of product that could also contains free radical scavengers. — 1

Total marks 6

10 Titanium has the highest strength-to-density ratio of any metallic element.

It can be produced in an industrial process known as the Kroll Process. Part of this process is shown.

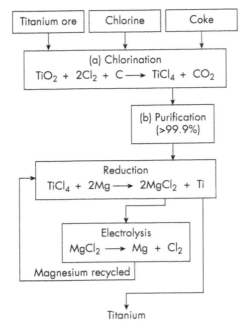

a Recycling is an important factor in the design of an industrial process. Magnesium is recycled in this process.

Draw an arrow onto the diagram to show how another chemical is recycled.

b 500 g of titanium chloride, $TiCl_4$, is produced in the chlorination stage.

  i Calculate the maximum mass, in g, of impurities that could be present in this sample.

  ii The titanium chloride is easily separated from impurities because it is volatile.

  Suggest the type of bonding present in titanium(IV) chloride.

iii If the titanium chloride produced after purification was 100% pure, calculate the mass, in grams, of titanium that would be produced if the percentage yield for the reduction process was 90%.

2

Total marks 5

11 The tables below show some of the first 20 elements from period 2 of the Periodic Table.

Table 1 gives covalent (atomic) radii in pm, and table 2 gives ionic radii.

Table 1 – Covalent Radii (pm)

Na	Mg	Al	Si	P	S	Cl
154	145	130	117	110	102	99

Table 2 – Ionic Radii (pm)

Na^+	Mg^{2+}	Al^{3+}	Si^{4+}	P^{3-}	S^{2-}	Cl^-
95	65	50	42	198	184	181

a Explain fully why the covalent radii of the elements decreases going from left to right across the periodic table.

2

b Suggest why the ions of sodium, magnesium and aluminium are significantly smaller than their atoms.

1

c The graph below shows the first ionisation enthalpies of the first 20 elements.

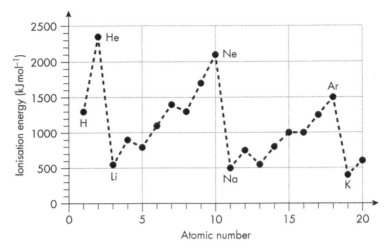

i Explain why the first ionisation enthalpy shows an increase going from sodium to argon.

ii Explain fully why the first ionisation enthalpy of argon is less than the first ionisation enthalpy of neon.

d A graph showing the ionisation enthalpies of oxygen is shown.

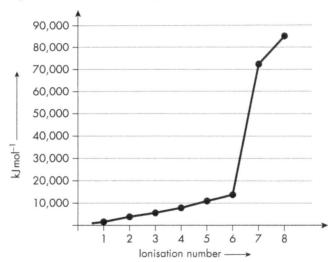

i Calculate the energy required, in kJ mol⁻¹, to convert one mole of oxygen atoms in one mole of oxygen ions with a charge of 2+

You may use your data booklet to help you. | **1**

ii Explain fully the increase between the 6th and 7th ionisation energies of oxygen. | **2**

Total marks | **9**

12 In ancient Egypt, the mummification process involved several steps.

One step involved adding compounds containing sodium to the body cavity, which removed water and also resulted in the saponification of fats, preventing decay.

a Saponification is the reaction of fats with an alkali to form soaps.

  i State another name for this reaction. | **1**

  ii Soaps can be used to remove grease.

  Explain fully how soaps are able to remove grease. | **2**

b When the body was dried, materials including cassia were added. Cassia contains the terpene, linalool and cinnamaldehyde.

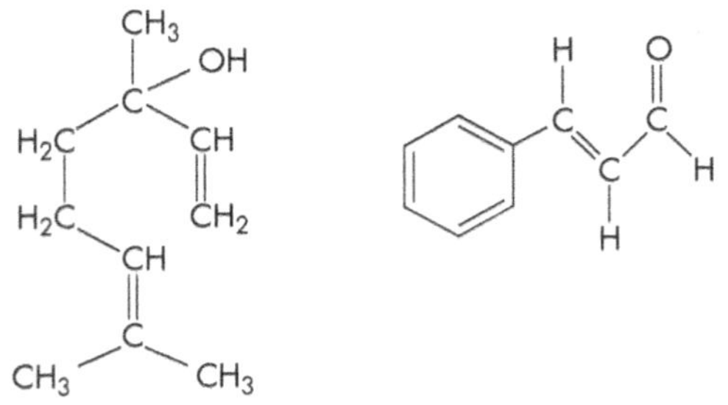

Linalool   Cinnamaldehyde

i Give the number of isoprene units contained within linalool. **1**

ii Write the systematic name of isoprene. **1**

iii Explain fully why cinnamaldehyde would react with acidified potassium dichromate but linalool would not. **2**

c Cedar oil, which is an essential oil, is also added to the body.

i State what is meant by the term *essential oil*. **1**

ii Give another use of essential oils. **1**

Total marks 9

13 Iron(II) sulfate can be taken as a dietary supplement.

The concentration of an iron(II) sulfate solution can be determined by performing a titration with acidified potassium permanganate.

$$5Fe^{2+}(aq) + MnO_4^-(aq) + 8H^+ \rightarrow 5Fe^{3+}(aq) + Mn^{2+}(aq) + 4H_2O(\ell)$$

a A standard solution of acidified potassium permanganate is required to perform the titration.

Describe fully how to prepare a standard solution of potassium permanganate.

2

b 10 cm³ iron(II) sulfate solution was titrated against 0.1 mol l⁻¹ acidified potassium permanganate solution and the results recorded in the table shown.

Titration	Volume of potassium permanganate (cm³)
1	9.1
2	8.8
3	8.7

Calculate the concentration, in mol l⁻¹, of the iron(II) sulfate solution.

Show your working clearly.

3

c During the reaction, the permanganate ion is reduced to manganese.

i State why no indicator is required for this titration.

1

ii Suggest why the permanganate solution must be acidified.

1

Extended response questions 129

d The recommended intake of iron per day for a 14-year-old girl is 14 mg per day.

Canned tuna provides 2.7 mg of iron per 85 g of tuna.

Calculate the percentage of the recommended intake that is provided by 100 g of tuna.

Show your working clearly.

2

Total marks 9

14 The rate of a chemical reaction is a measure of how quickly reactants are converted into products.

Using your knowledge of chemistry, comment on what influences the rate at which the reactants are converted into products.

3

Notes

Notes

Notes

Graph paper

Graph paper

Graph paper

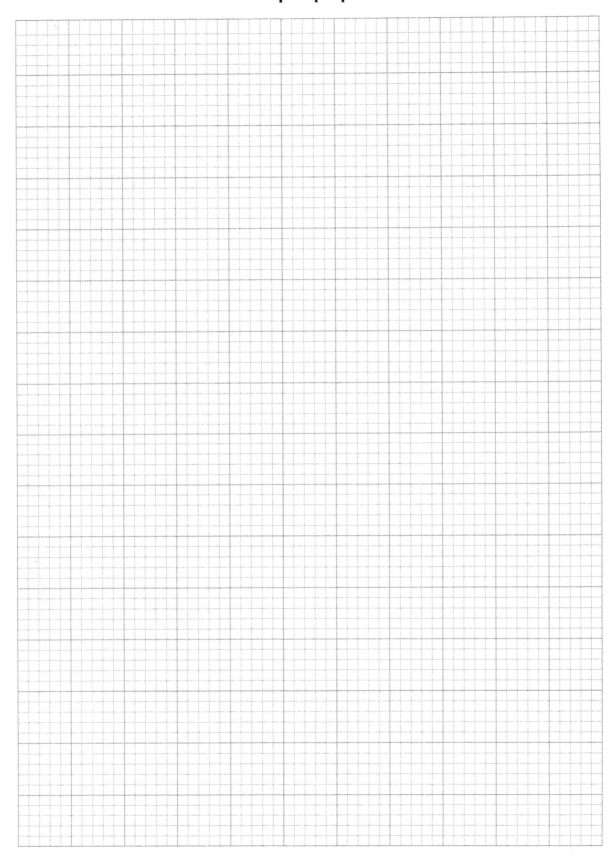